FORSAKEN EARTH

THE ONGOING MASS EXTINCTION

PAUL SEQUOIA RAUCH

ISBN: 978-1-4834-5449-8 (sc)
ISBN: 978-1-4834-5448-1 (e)

Lulu Publishing Services rev. date: 07/26/2016

Dedicated as a record

for the future

and

a hope for

the present.

TABLE OF CONTENTS

List of Animals on Cover:

All critically endangered - likely less than five years left of existence (www.iucnredlist.org). *Lipotes vexillifer,* "The Goddess of the Yangtze," is now declared extinct.

Dama Gazelle, *Nanger dama*; Northern White-cheeked Gibbon, *Nomascus leucogenys*; Red Ruffed Lemur, *Varecia rubra*; Javan Blue-banded Kingfisher, *Alcedo euryzona*; Western Gorilla, *Gorilla gorilla*; Sumatran Orangutan, *Pongo abelii*; Cotton-headed Tamarin, *Saguinus Oedipus*; Yangtze River Dolphin, *Lipotes vexillifer*; Hawksbill Turtle, *Eretmochelys imbricata*; Elkhorn Coral, *Acropora palmate*; Vaquita Whale, *Phocoena sinus*; Southern Bluefin Tuna, *Thunnus maccoyii*.

INTRODUCTION

Fortunate to be born to the temperate rainforest of the majestic, shimmering U.S. Pacific Northwest, the forest became my sanctuary; nature became my healer, teacher, and protector at a very young age. Before ten, my grandfather recognized this and gave me a subscription to a wildlife magazine for an amazing birthday present, so I have been aware of environmental issues all my life. As a teen wanting to live in the wilderness, wild edible plants became my study, and at sixteen I took a 24-day Outward Bound course in the Washington North Cascades wilderness areas; I was in heaven like Brer Rabbit in a briar patch and didn't want to come back. I returned throughout my teens and twenties to this paradise of virgin mountain wilderness flowing with Earth's purest waters, and I often scrambled off-trail, bivouacking for weeks and months without even a tent, solo. Solo because no one else had the same passion I had; it was a calling so strong, I could not help but choose it over the standard path of college or career. I cherish this extraordinary experience to this day, the profound impact and beauty of it lasting a lifetime.

As a kid I'd wanted to be an astronaut, so I had long imagined the Earth from space many times before the famous Earthrise photograph from the Moon was published by NASA. In the mountains, I profoundly felt stark reality under the bright Milky Way; I felt the rotation of our Earth and the orbital dimensions of the visible planets placed in our solar system, also watching and understanding their retrograde loops; and I felt our solar system's dimensional placement in the reality of our galaxy. With this perception and my love of all life on this magnificent, miraculous Earth, I have been filled with amazement, and I cannot accept or sit by and allow life to be enormously desecrated by a mass extinction mindlessly and heartlessly caused by our own species.

Writing *Forsaken Earth* became as unavoidable of a calling as the more than ten years living in the wilderness every summer had been in my youth; both were something I *had to* do. Everything pertaining to the ultimate and most critical issue on Earth - the ongoing mass extinction of species - is addressed in this one iconoclastic volume. Largely unrecognized and denied, this extinction event is occurring right now and rapidly accelerating during everyone's current lifetime right in front of our eyes. It is with crucial intent that our current scenario must be identified and diagnosed for what it is. Integrating hundreds of references from over one-hundred nonfiction books and documentaries on the subject from some of the most renowned authors, including Pulitzer Prize winners, *Forsaken Earth* clearly defines the causes and details of the human-caused ongoing mass extinction of species we all must face.

Most abhor the ineffably tragic extinction that causes so much beauty and life of this gorgeous, one-of-a-kind planet Earth to be lost, and that's just from what little they hear about it and then brush away from their minds. But deep down, part of our heart and soul dies, even within the most unaware, with each species utterly and forever vanished and with each ecosystem that becomes destroyed and deserted. Though it is currently where we're headed, to not settle for this as our accepted and inevitable common destiny requires us to become more aware of it and become involved in actively creating a truly new world.

Life on our planet desperately needs our species to experience a great awakening *now*. I imagine the feeling and possibility of our species' humble triumph, as people all over the world finally comprehend what's at stake and finally care enough, when true realization hits, to take action to heal the grave imbalances we've caused our Earth and one another. Writing *Forsaken Earth* has been an act of optimism towards this result.

The most worthy purpose of all awaits everyone in becoming part of our own and our Earth's healing. Our great awakening will be an exultant one and cannot come too early; it is the Great Work of our times. It is a shared destiny which could also finally unite all people through the heart and force different decisions of the corporate-controlled governments, a true paradigm shift. With a new "spirit of the times," it may be easier than we currently believe, especially as we increasingly transform the old and toxic inertia, the accepted torpor of "the way things are." Soon having phones that translate languages in real-time can only help create a global voice that effectively rejects "the way things are." As a beaming Kenyan taxi driver said on a recent TV news story about smartphone technology creating a new monetary currency, "Anything is possible in the new world."

We are at a time in history where there is no longer any time to lose, a time when the deadly status quo is *inexcusable* and worsening. If you think someone else will take care of everything, reconsider that those of the caring on this issue are, as yet, far outnumbered by the uncaring or desperate. But the lights can go on instantly for anyone, and such light and all life of Earth needs *you*. True purpose and fulfillment awaits you as an active participant in the movement to preserve all the amazing and indispensable myriad lifeforms of Earth, once known as Paradise, before half of Earth's species are truly and permanently extinguished.

URGENCY, PART I

O N A GLOBAL SCALE, our world is currently losing species to
extinction at a rate of over two per hour, twenty-four hours a day,
every day and night, and the rate is increasing at all times without
pause. Many scientists are stating that 70,000 species are already going
extinct each year.[1] Others say much more, but using the most *conservative
estimates* of correlated sciences, today's rate of species extinctions is about
20,000 annually[2], which is one every twenty-six minutes. No matter the
approximations, the problem is already incomprehensibly tragic and
unacceptable. "The labor and care expended over some billions of years
and untold billions of experiments to bring forth such a gorgeous Earth
is all being negated within less than a century for what we consider
'progress' toward a better life in a better world."[3] The full extent of
species loss is expected to happen between 35-85 years (2050-2100).
Without changing our ways, by the sixty-year average of this likelihood
(2075), fifty percent of our planet's unique species will be gone from our
world, from the Universe, from Creation, forever.

The ongoing mass extinction is 100 percent anthropogenic; the
world would not be currently suffering a mass extinction if not for human
causation. While the world focuses on many rivals being the number one
concern about our Earth, such as climate change or chemical pollution,
the Human-caused Ongoing Mass Extinction (HOME) is the ultimate
result of them all. More than just crimes against humanity or crimes
against nature, "species extinctions... [are] the greatest crime against
creation."[4]

The most commonly agreed-upon number of species that exist on
Earth is 9 million. This number does not include bacteria but does
include our estimations of centinelan species (those not yet known).

Before humans, one species had naturally gone extinct per million per year, known as the background rate, as discovered by paleontology and other sciences. This means nine species would naturally be expected to go extinct per year (one every forty days), generally replaced by a similar amount of new species through speciation, both aspects of natural selection. Now, however, "[t]he rate of extinction of animal and plant species is estimated to be at least several hundred times greater today than the natural rate."[5]

Our ideology determines our beliefs, policies, and actions; if no change occurs with our ideology, even with technological advances, the average rate of the extinction of species equates to one every seven minutes. This is the average starting *now* and factored for sixty years. By then humans will have caused the extinction of 4,500,000 species, or 75,000 species per year, 206 per day, or nine per hour twenty-four hours a day. Without changing our "business as usual," for the rest of your life an average of seventy species go extinct as you work your eight-hour shift or every night while you sleep, or about thirty while you watch an American football game, or more than a dozen while you go to the movies, or nine for each hour you watch TV, and so on. This is a linear calculation, but nature's events will not be as linear as our rationality.

As the fabric of life disintegrates, trophic cascades (the domino effect of food-tier losses) are caused; for instance, with over 26,000 plants in the world now threatened with extinction, the animals dependent upon them for food are also directly threatened and so on. We are in an ongoing mass extinction, headed towards a collapse of species, and like a berg calving off an ice shelf or a sandbank undercut by a river, at a certain point the number of extinctions will explode. If we don't make drastic changes to heal our relationship with the nature of Earth, when the collapse of species comes, the average rate of extinction will rise to a percentage in the hundreds of thousands higher than the normal background rate.

AMPHIBIANS

"[A]mphibians… are the ancestors of all living
terrestrial vertebrates, including humans."[1]

N O OTHER CLASS OF animals within the animal kingdom is a better barometer of the anthropogenic, environmentally despoiled state of the world than amphibians. It is gravely alarming that one-third of approximately 6,000 known species are threatened with extinction and are going fast. Being cold-blooded, amphibians are more susceptible to the ongoing erratic climatic changes and are particularly vulnerable to drought. Being water-born, they are also defenseless against the proliferative use of manmade chemicals that continually end up in their waters. Although a huge factor is habitat loss, even worse is that many amphibian populations have declined in areas where their habitats are intact. This tragedy of amphibian loss has been increasing so much since the 1980's that scientists refer to it as the Global Amphibian Decline, and there is no evidence in the fossil record that any extinction of amphibians has ever occurred so suddenly and to such extreme proportions.

"Amphibians emerged at a time when all the land on earth was part of a single expanse known as Pangaea."[2] Towards the end of the Devonian Period in the middle of the Paleozoic Era (370 million years ago), DNA branched life into a new form of animal in pursuit of the first plant forms that migrated onto land out of the swamps and oceans. This entailed DNA advancing from fish by losing scales and growing legs, beginning with species such as the *Ichthyostega*. This migration of life resulted in the current form of amphibians by 350 million Before

Present (BP), about 120 million years before the dinosaurs' 165-million-year reign on Earth, and 210 million years before flowering plants.

Amphibians have a global presence with very specialized niches in every biome on Earth except for the Polar Regions, the upper reaches of mountains, and salt water. Both aquatic and terrestrial, some amphibians retain their gills, some have primordial lung-like sacs, and others respire solely through their skin. Today, almost 90 percent of amphibian species are frogs and toads, the rest include salamanders, newts, and caecilians.

Although CO_2-caused acidification helps mercury accumulate in their bodies, amphibians are threatened by many events other than the effects of climate change. Roundup and atrazine are omnipresent in their environment, poisoning them as they absorb these chemicals through their skin and consume other chemicals in the bugs they eat, causing gross mutations. Places such as Haiti have demolished 99 percent of their formerly-lush, amphibian forest habitats that also polluted all the amphibians' fresh surface waters with silt, both events critically endangering the endemic and other species there. Fragmentation of amphibians' habitats leaves them prone to more loss of numbers. Cattle grazing ruins springs, lakes, and streams with direct input of protozoa-filled feces and the trampling of everything. Amphibians are very susceptible to our world's increased exposure to ultraviolet B radiation as a result of the continual use of manmade chemicals that still deplete our stratosphere's ozone layer. Lastly, similar to bats contracting their lethally contagious white-nose fungal disease, a present syndrome first spread by humans recreationally entering their caves, amphibians also suffer from a lethally contagious fungal disease called chytridiomycosis, or chytrid, which is now global and causes them to go into convulsions and die of heart failure. The spread of this fungus began in the 1930's with the use of African Clawed Frogs from South Africa for women's pregnancy tests around the world, and it further spread through increased trade, development, and large agriculture.

As much as the threat of global warming attracts the attention of the world, the loss of massive numbers of amphibians now actually occurring should be of even greater concern. Losing increasing numbers of these animals is indicative that nature is in an unprecedented imbalance;

amphibians are the first vertebrates to migrate from the oceans to inhabit land and have been nearly omnipresent on land ever since, living their happy little lives as major predators of insects. With the ensuing rise of insect populations, the dominant culture (DC) will dispense even more chemicals into our environment, which was the major cause of the amphibians' demise in the first place.

We are losing amphibians to eternal extinction and replacing them with a world saturated with manmade chemicals, both firsts in Earth's entire history.

CHEMICALS

IN 1962, RACHAEL CARSON wrote her prophetic book, *Silent Spring*, correctly warning about the lethal effects pesticide use has on wildlife and humans. Her information led to new laws being passed banning the use of some chemicals in the U.S. such as DDT (an organic insecticide/pesticide). It was a bump in the road for the chemical-producing companies, however, which have massively amplified production since, plus adding new chemicals every year without objective safety tests. The world now uses billions of pounds every year of over 80,000 types of chemicals: pesticides, herbicides, fungicides, fertilizers, and many more. A few are banned in the U.S. but approved in other countries that export their tainted food products, and others, back into the U.S.

Traces of hundreds of persistent organic pollutants (POPs) that have only existed within the past three human generations are found in human tissue. These include DDT, PCBs, dioxins. polybrominated diphenyl ethers (PBDEs), perfluoroalkyl substances, stain repellants, lubricants, and more; all are bioaccumulative, ending up in the highest tiers of the food chain, such as in polar bears, some of whom are becoming hermaphroditic as a result.

Chemicals kill tens of millions of fish, birds, and amphibians per year just in the U.S. Atrazine is an herbicide produced mostly by a company named Syngenta and is banned in several European countries, as it is known to kill amphibians. Traces of atrazine are found in all the rainwater of the U.S., a country which still uses over 75 million pounds of it per year.

Glyphosate, another herbicidal chemical known as Roundup, is now made by companies other than Monsanto under generic brand names and is massively produced in China and other countries. Monsanto also invented Agent Orange (2,4-D), DDT, and PCB. The use of glyphosate was about 150 million pounds in 1999, but it now has an annual multi-billion-dollar international market, soon to have a yearly use of almost 4 billion pounds. Largely sprayed on genetically-modified U.S. and Brazilian soybean crops and others around the world, it is also widely overused by the public; whole cities can reek of it in the spring and summer.

"In the United States alone, we now use more than 18,000 different pesticides, a dramatic jump from the 200 that were used in the early 1960s. ...50 years ago we were using 400 million pounds of pesticides a year, and today we're using more than 4.5 *billion* pounds a year. In that same time span, the industrial chemical business has grown from a $2 billion industry to a $635 billion industry."[1] One of the pesticides being dispersed into the global environment is methyl bromide, which is still heavily used in the U.S. and is known to damage the ozone layer. "A graph showing the consumption of these chemicals [not just pesticides] in the United States alone, if one inch of rise represents 100 million kilograms of chemicals consumed in a year, looks as follows: If we start at 1920, the line begins 3/4 of an inch from the bottom axis.... Then between 1935 and 1940 it rises to 100 inches, and between 1945 and 1995 it rockets to the height of a 9-story building."[2] From 1935 to 2015 is just eighty years, one human lifespan, an irreversible experiment in spreading toxic chemicals globally in a blip of time since the beginning of agriculture, a "gigasecond" of relative time, disastrously affecting all life on our planet. There can be no doubt or debate about chemical pollution being anthropogenic, and it is already decimating amphibians for the first time since their 350-million-year existence.

In total, the chemical industry has a global market in the upper hundreds of billions of dollars in annual sales. Monsanto is not even in the top ten companies: Dow Chemical, ExxonMobil, DuPont (USA), BASF (Germany), Sinopec (China), and Shell (Netherlands) are. Some of these companies' cancerous chemical inventions are used for fracking and flow into the millions-of-years-old, pristine, underground aquifers of our Earth. The danger of inundating our Earth with manmade chemicals by far supersedes anthropogenic climate change in its immediate, ongoing effect upon the entire gene pool of life.

URGENCY, PART II

WE ENTERED INTO THE beginning of the Human Exponential Event (HEE) - the exponential growth and activities of the human population - when the Industrial Revolution took off as the result of our ability to pump oil from the ground in the 1850s. The use of oil finally allowed our species to improve our ability to feed ourselves so much that domination of the Earth truly took hold.

The extinctions of non-human species started off marginally with the demise of endemic species on the world's islands, including the "island continent" of Australia, all with the highest extinction rates on Earth. These rates are now occurring on all the continents on Earth; in the past twenty years, mainland extinctions have equaled those previously of islands. Even on and around Antarctica, a continent with no development or human population, thirteen of the world's eighteen penguin species, which have been around for 55 million years, are listed as threatened with extinction. Generally, wildlife that hasn't been made extinct yet has been decimated by 95 percent since the 1850s, species such as orangutans, bison, whales, elephants, tigers, Bluefin tuna, rhinos, cod, lemurs, chameleons, prairie dogs, turtles, wolves, chimpanzees, gorillas, manatees, gibbons, certain bears, birds, seals, sharks, and *thousands* more, as well as redwoods, sequoias, mahogany, teak, koa trees, and countless more plant species. Humans have stopped the process of evolution for larger vertebrates of the animal kingdom, and we are annihilating species and entire ecosystems that may never be salvaged.

Providing for human prosperity by extinguishing half the life of our planet and its ecosystems is an inexpressibly foolish and heartbreaking oxymoron. Our mindless abolition of Earth's animate and inanimate resources has pushed our planet into an ecological disaster, and what impact humans have upon the Earth is swiftly becoming more irreversible every day.

Because humans have become such a potent geological force, many geologists state we have moved out of the current Holocene geological epoch (the 11,500 years since the end of the last Ice Age) and have entered into a new, Anthropocene epoch. Never has a species (ours) caused and defined a geological epoch. And never has a species caused a mass extinction, the last phase of which has begun and is expected to surpass the meteoric collision with Earth 65 million years ago that caused the extinction of dinosaurs. The ongoing extinction is the result of our species changing the ecosystems of Earth more rapidly and more destructively than ever in oceanic life's 3.4-billion-year history and terrestrial life's 450-million-year history. Even with the dinosaurs' demise, many species survived unharmed and their DNA was left uncorrupted, but the current HEE is impacting all species everywhere in very novel ways, such as with gene-destroying chemicals, radiation, and continued ozone destruction.

Any people alive now and onward are facing a mass extinction of species that is altering the course of life on Earth forever. The environmental damage humans have caused will continue for several millennia even if humans abruptly vanished from Earth. Our world is full of species already known as the living dead (the last individuals of a species still to exist but who survive under circumstances too formidable to generate recovery). To stop this syndrome we need to put our full intention on changing everything, most of all our entire ideological paradigm. Although it will not be the complete end of life on Earth, nor likely of humans, "[w]e are truly on the brink of a catastrophe. In such a situation, radical change… is the only hope for survival."[1]

We do know that the mass extinction has already begun, and that it's the ultimate result of the HEE. For instance, human beings today use an average of fifty times more resources than did the Stone Age foragers before agriculture. This equates to the average lifestyle of the world's present population of 7.4 billion equaling over 350 billion Paleolithic foragers.[2] The average is greatly heightened with the lifestyles of North Americans, Western Europeans, Japanese, and Australians who consume about thirty times more resources and produce that much more waste than an average Kenyan. Hence, in about two years, over-consumers of the world go through more resources than a Kenyan does in their entire life.[3] Our Earth is already unable to sustain our current populations and standards of living; so far, the environmental debt we are creating is being paid by greatly increasing extinctions of species and loss of ecosystems.

"[T]he impetus to preserve biodiversity should be our greatest – and most urgent – call to arms."[4] All other problems from the HEE on Earth are the cause of this greatest problem – the loss of biodiversity - that far surpasses all others combined, and yet, most people remain unaware or treat it as if it were of no importance. However, it is the most rueful catastrophe that could ever happen. "The central reality of our era is extinction. Nothing is more important."[5] If we ever realized the peril we've caused, we would not have allowed this to have begun, would not remain in massive denial about it, and certainly would have admitted and corrected our errors by now. The largest problem on Earth remains unrealized, denied, and grossly minimized. It has been ignored with irrelevance and apathy for at least decades.

Instead, the DC, the controlling power in the world for thousands of years, is culminating in ruining the natural world. To name a few, the dominant ideology of the culture (DIC) is ruining our oceans with acidification, overfishing, plastic, bottom trawling, and radioactive waste; our atmosphere with CO_2, methane, and ozone-depleting chemicals; our underground aquifers, millions of years pure, with fracking chemicals and agricultural depletion; our forests with total destruction replaced with genetically modified organism (GMO) tree farms, cattle feed lots, mining, and other uses; our rangelands with the severe devastation that comes with the overgrazing of domestic ruminants; our fresh water systems with dams, wetlands draining, waste, and thousands of chemicals to the point where fish and frogs can't even survive; our mountains and other lands with giant, industrial mining for coal, minerals, and shale oil; and the elimination of keystone species, completely disrupting even the scarce, protected, wilder habitats.

Misguidedly, the DC believes we'll function just fine without an intact natural environment and the permanent fatality of half the world's species, and the dominant ideology (DI) can't stop the destruction it's causing with the current and long-standing economic conviction it has. Business and politics are unable to function in real terms but rather in short-term, bottom-line data and the necessary denial of science in order to continue operations; as a result, thoughtlessly cutting wildlife populations to the bone to minimally comply with endangered species laws (if at all) has become acceptable. The dominant cultural ideology (DCI) is without concern for the impact it's causing, which is done for enormous profits to fill the gluttonous extravagance some narcissists of our species lavish upon themselves for so-called status. On the other end of the spectrum, billions of people, just as destructively, are desperately trying to simply meet their daily needs of survival.

It's indefensible that any species are already going extinct. We mistake it as normal and acceptable, the price we pay to increase human progress. However, the price of our living on Earth must require that other species do not go extinct, for doing so would force us to keep our Earth healthy. No countries are willing to pay such a price, a price to restore wildlife and wildlands truly protected for all time that never should have been decimated in the first place.

We must consciously evolve: there is nothing left to do but heal ourselves and the Earth. The HOME, the loss of our Earth's biodiversity, is the greatest crisis in our world, and "[t]he measure of change has gone from the millennium to the decade."[6] The warnings have been sounded for centuries, first by cultures that are still completely disregarded. Now, even intelligent and dedicated military men of countries around the world conclude that "the environment... poses the greatest threat to the security and to the survival of the human race."[7] Even the cause of the dinosaurs' demise purportedly took longer than the one currently escalating. Furthermore, this extinction will not stop until humans quit being the cause, either voluntarily or not. And we no longer have any time to lose: the time is now upon us.

Since the passage of the U.S. Endangered Species Act in 1973, "more than 1200 species in the United States have been listed as either endangered or threatened,"[8] yet several thousand more qualify for listing but are blocked by corporate interests. For instance, over "4600 plant species... known to exist in the United States, [are] in danger of becoming extinct."[9] And that's just plants. All around the world, "...the immense storehouse of genetic diversity in the oceans – particularly the nearshore areas – is being destroyed by human developments faster than it can be cataloged."[10] "The destruction of 70 percent of the natural world in 30 years [and] mass extinction of species... is forecast... by 1,100 scientists...." - *The Guardian*, May 23, 2002.[11]

Currently, the "vanishing of tropical habitat... [puts] the current global extinction rate as high as 1000 times the 'normal' background rate."[12] A career biologist of the tropics, John Terborgh, states in his book, *Requiem for Nature*, "...a lifetime of roaming the Tropics in quest of unspoiled nature has dispelled any complacency and replaced it with panic."[13] It's not just the wild going extinct, either: "...30,000 vegetable varieties were lost during the twentieth century, and [one is now] going extinct every six hours."[14]

Evidence that the ongoing mass extinction is already taking place is very clear, but it is getting very late for humanity to become aware and convinced of it enough to save the ecosystems of endangered species. In recent years, primary environmental organizations have realized that it is and will be impossible to save all endangered species, and they have already adopted a strategy of triage. With the environmental devastation of the countries of West Africa having become so dire, it has become the first global area to be written off as a waste of precious conservation resources, since attempting to restore environments and species elsewhere would at least have some odds for success. Developers love it because there's nothing to save and thus nothing to legally impede them, and environmentalists have better publicity by focusing on ecosystems and species they at least have a chance of saving.[15]

Since about 1945, humans have restructured our planet's ecosystems faster than any other time on Earth; we have created unprecedented, global, life-threatening circumstances for all life on this planet, and there is no second chance to get it right. All these past eighty years or less only 20 percent of humanity reaped the rewards of industrialization. Now, as 80 percent of the world approaches industrialization (with triple the amount of people since 1930, that's a 1200 percent increase over what is already unsustainable, with billions more people coming), without a completely new paradigm, half of non-human species will vanish forever from the Earth with the ensuing annihilation of all ecosystems. Our critical obligation is to heal, restore, and rejuvenate our natural world.

Twenty-five acres in parts of Indonesia (and other tropical forests) once "contained as many different tree species as are native to all of North America. This wealth of biodiversity may disappear in smoke before we even know what we have lost forever."[16] "The library of life is burning and we do not even know the titles of the books." – Gro Brundtland, Former Prime Minister of Norway.[17]

We are all Nero; instead of "fiddling" while all of Rome burned, we are now collectively creating, allowing, and watching the entire Earth become ravaged and forsaken. Our paradigm, the ideology under which we live, is completely faulty, broken. All the proof we need is already here: our anthropogenic mass extinction of species, an escalating event from which we are not immune.

CORAL

CORAL ARE CNIDARIANS, PART of the animal kingdom in the same phyla as jellyfish and sea anemones. Their first ancestors came into existence with the onset of the Cambrian period up to 590 million BP. The animal kingdom had started with simple oceanic animal forms without skeletons about 210 million years prior, at 800 million BP. This followed a span of life on Earth that consisted solely of one-celled microorganisms for about 2 billion years. Thus, after almost 3 billion years of life, coral were an evolutionary development that joined the relatively new, skeleton-less Kingdom of Animalae. The hard coral we know today came into existence during the Triassic period 200 million years ago, and the reefs we have now took 50 million years to grow into the healthy, pristine status they held before our enormous human impact.

Though coral exists throughout most of the oceanic world, the spectacular and exotic reef-building coral environment is only present between the latitudes of 30°N to 30°S in subtropical and tropical waters with temperatures between 64-91°F. It is the change in temperature which determines the boundaries between coral reefs and the cooler kelp forests. Coral reefs are known as the rainforests of the ocean; coral are keystone species, providing habitat for hundreds of thousands of other species. Up to 40 percent of all marine fish species live within coral reefs. Coral use carbon to form their shells and ultimately contribute to form limestone, which constitutes the largest biologically-made formations on Earth.

Less than one percent of the world's ocean habitats are designated as protected similar to a terrestrial National Park or wilderness area. Only 1.7 percent more is "protected," meaninglessly, as these areas have zero effective management or enforcement and are open to commercial fishing. The Great Barrier Reef, Australia, was made into a marine park in 1975 and acquired designated protection as part of the UNESCO World Heritage List in October 1981. It is the largest protected coral habitat in the world and has a length of 1,250 miles, equaling that of the three-state U.S. Pacific coastline. It contained 2,900 individual reefs, which supported at least 1,500 species of fish, 4,000 types of mollusks, and 450 kinds of corals. However, "coral cover in the Great Barrier Reef has declined by fifty percent just in the last thirty years."[1]

The Earth's hotspot of marine biodiversity exists just north of Australia in what is called the Coral Triangle, the Southeast Asian archipelago, around the Philippine and Indonesian islands and Borneo, an area which also includes mangrove forests full of endemic life. The Coral Triangle has the greatest assortment and amount of marine species found anywhere on Earth.

As up to 75 percent of the world's human population will live within one-hundred miles of a coastline within four years, and a huge amount of this population is in the Coral Triangle, this gorgeous marine paradise will experience even greater catastrophic loss of species in its coral reef and mangrove habitats than it already has. This takes form in destructive fishing methods, which use dynamite and sodium cyanide; in sewage disposal; in oil drilling; in tourist-related activities; in overfishing, especially of herbivorous fish, causing a severe excess of algae; and in erosion from tropical rainforest logging, causing eutrophication (oxygen loss) by depositing great amounts of silt into the reefs, which kills all the algae. Coral and algae have a symbiotic relationship: too much algae smothers coral, and too little algae leaves coral to starve and lose its color, known as bleached coral.

Noticeable outbreaks of the poisonous crown-of-thorns starfish (*Acanthaser planci*), which feast on reef coral and have been termed a plague, started occurring in the 1960's when the anthropogenic imbalances mentioned started having their effect throughout the Pacific and other regions. The affected areas stretch from the Red Sea, East Africa, the Great Barrier Reef, the Coral Triangle, and Hawaii to the west coast of South America. The crown-of-thorns consumed up to half of the reefs in decades-long cycles.[2]

All of that is enough to cause catastrophic destruction, but then there's climate change, the single greatest hazard to the coral ecosystem. The ocean is getting more acidic, as it absorbs 30 percent of the CO_2 human activity is emitting; the ocean is also getting warmer, as a result of the troposphere heating up from having 64 percent more CO_2 in it (so far) than it has in hundreds of thousands of years or more. Thermal tolerance varies by coral species and physical location; for instance, reefs in the Florida Keys grow at 64°F, and healthy colonies exist in the Coral Triangle and the Persian Gulf at 91°F, but the ocean only needs to be warmed for several days by 2°F more to cause the algae that supports the coral to disappear or die, causing coral bleaching.

In the late 1980's, an International Union for the Conservation of Nature (IUCN) survey estimated there were 234,000 square miles left of live coral-reef ecosystems. By 2003, higher oceanic temperatures had caused 70-90 percent of corals from Costa Rica to Ecuador to expire, including 95 percent of corals in the Galapagos, with no signs of recovery. "[T]here are no live corals left in the Maldive Islands."[3] The Maldives are comprised of 2,000 small islands in the Indian Ocean and are already being threatened with inundation from rising sea levels. Presently, up to 90 percent of the reefs in all of Southeast Asia approach highly threatened status from anthropogenic undertakings; the numbers in the Philippines reach almost 100 percent along with over two-thirds of the mangrove forests having been already destroyed; corals in the Gulf of Mexico and the Caribbean islands have disappeared by 80 percent; and half the coral in Australia's Great Barrier Reef is dead, with more severe abuse occurring from coal mining exportation as a result of dredging and dumping for giant port expansions.

"[T]he proportion of coral species ranked as 'threatened'… exceeds 'that of most terrestrial animal groups apart from amphibians,'" and, "…within the next fifty years… 'all coral reefs will cease to grow and start to dissolve,'" due to ocean acidification, and "[i]t is likely that reefs will be the first major ecosystem in the modern era to become ecologically extinct."[4] With their death, even the term "catastrophic loss" is an understatement.

BACTERIA

T HE KINGDOM OF MONERA, the prokaryotic kingdom, consists of the miraculous creation of the first lifeforms to exist on Earth. Comprised solely of bacteria, these one-celled organisms have ruled our world for 3.4 billion years, and they were the only lifeforms on our planet for at least the first billion of those years. All prokaryotes reproduce asexually and are the only lifeforms on Earth lacking a nucleus within their cell to contain and protect their chromosomes, their DNA.

Bacteria are amazing; without them, the rest of life never would have begun or led to our human existence. Our lives still depend on them; without them we would perish. Yes, there are some that cause disease, but out of the over 10,000 species that have been named and described, only a little over 500 are harmful to vertebrates; of those, only one-third affect humans. When one considers that less than 1 percent of all bacteria have been studied, leaving at least a million more types that still remain unrecognized, the percent of bacteria that is harmful to humans is infinitesimal. It has been calculated there are more individual bacteria in existence on Earth than there are stars in all the galaxies in the entire Universe - and there are about 100 billion stars in each of the estimated 100 billion galaxies.

All bacteria are so small that they are invisible without a microscope. Their size helps give rise to their great numbers. For instance, it is estimated there are over 6 billion bacteria in your mouth; this number consists of 700 different bacterial species. Our intestines hold up to 100 trillion prokaryotes, plus they are in and on every other part of our bodies; this means that by number, not size, our bodies consist of 90 percent bacteria. Soils contain millions of bacteria per cubic meter or thousands of species in just one gram. Additionally, there is an entirely different prokaryotic world that fills at least a two-mile depth of earth: subterranean lithoautotrophic microbial ecosystems, or slimes, as they're known mnemonically. Slimes are mostly anaerobic bacteria and microscopic fungi that probably weigh more than all other terrestrial lifeforms combined.[1] Prokaryotes are omnipresent, from deep ocean waters and subterranean earth to heights in the atmosphere; they are also the first to inhabit or reoccupy land that has been freshly created or made lifeless by fire.

Without bacteria, Earth would not have its existing atmosphere. For eons, these organisms have produced, reduced, or converted all of Earth's reactive atmospheric gases, gases such as oxygen, nitrogen, carbon dioxide, hydrogen, methane, ammonia, several sulfurous gases, and many more.[2] Prokaryotes also break down into atoms all that has died. Cyanobacteria were the first lifeforms to develop photosynthesis, which oxygenated Earth's atmosphere. From the waters of the ancient world, this feat took half a billion years, one atom at a time. Wild colonies of these bacteria still exist in Western Australia, called stromatolites, and other various types are grown commercially as health foods, such as spirulina and chlorella.

Moneran also feed all of life, as they are the beginning of both terrestrial and oceanic food chains. Plants would not exist without bacteria, as prokaryotes provide the needed soil nutrients as well as the nutrients plants transpire from the air.

That the Kingdom of Monera still rules our world is exemplified by how it changed human history with the death and migration of millions of Germans and Irish with the potato blight of the 1800's, or the death of millions from the Black Plague in the 1300's, or how it has increased our technological advances with the immense amount of time and resources spent researching how to harness the power this kingdom of life has over our lives.

COSMOS

UNTIL ASTRONOMER EDWIN HUBBLE proved otherwise in 1925, it was believed our galaxy *was* the entire Universe. Before then the scientists of our world could not even conceive of anything existing outside of our own enormous galaxy, a galaxy 100,000 light years in diameter, one light year being a distance of 6 trillion miles. Ninety years later and to very simply explain, astrophysicists say all the mass of our unimaginably vast Universe was contained in the size of a single atom before the Big Bang[1,2], "an infinitely dense state,"[3] and the size of the Universe, though still expanding, was literally created in an instant, which is all still only a theory. Scientists say the instant expansion of the Universe happened because the speed-of-light limitation only applies to particles (mass) and not to the expansion of space, because space is not made of particles. Additionally, galaxies are not moving away from one another as previously thought; rather, the space between them is stretching, and the stretching expansion, caused by energy we cannot detect, is accelerating.[4,5]

We thus have the age of the Universe at 13.69 billion years, the time it has taken light to travel the distance from the Big Bang event to us, but the space or size of the Universe now extends 44 billion light years from Earth.[6] Both of these measurements are of radius, so the size is doubled for diameter, squared for area, then cubed and rounded to a sphere shape for volume, culminating in an ever-increasing number with dozens of zeroes after it. This space is filled with at least one hundred billion galaxies of sizes similar to ours. Astronomers have estimated a count of 10,000 galaxies in a space one-tenth the diameter of the moon; such galaxies extend out in every direction. A famous photo taken twenty years ago by the Hubble Space Telescope, called the

Hubble Deep Field, exposes hundreds of galaxies from about four billion light years away; "[t]he area of the [photo] equals about one-hundredth the area of the full moon, or a grain of sand held at arm's length."[7] There is light out there all around us of galaxies and quasars, light that's too far away to see without immensely powerful telescopes staring at the same spot over several nights. The Universe also measures and basically looks the same from any other galaxy; a central orientation exists equidistantly for all galaxies within our Big Bang sphere of light.[8]

At least on paper, science has established a miracle with its mathematical "proof" that our Universe was all created in an instant from invisibility out of nothingness, and the Universe, the vast majority of which consists mostly of dark matter and dark energy, which is also as yet utterly invisible, is defying gravity and accelerating the stretching of space. "The current theory is that the universe expanded extremely fast [in a millisecond] after the big bang and then started to slow down due to the action of gravity. But after about eight or nine billion years, dark energy won its battle to overcome gravity and started to drive the acceleration of the expansion that we now observe."[9]

Regardless, what's relative is that our Earth does exist within our solar system at 27,000 light years from the center of our galaxy. With a galactic orbital circumference of 1.017 quintillion (a billion billion) miles, our solar system takes 240 million years to complete one orbit, called a galactic year (GY). Thus, Earth's age of 4.6 billion years (one-third the age of our Universe) is comprised of over nineteen galactic orbits, making our present galactic date 19.17 GY. Accordingly, with our solar system speeding at about 500,000 miles per hour around our galaxy, the Earth has traveled a total of 19.5 quintillion miles, a distance of 3,250,000 light years.

Distant supernovas and other cosmic explosions created the needed atomic elements that gravitated together to finally form our Earth. For the first 2.92 galactic orbits (700 million years), our Earth was bombarded with planetesimals containing enough ice to supply our future oceans; however, Earth was completely molten, and all the water was gaseous until all the heavier elements, such as iron, sank to the core of the Earth and the planet cooled enough for the rains to precipitate like waterfalls from the sky.

Once the oceans were formed, it took another 500 million years for the first lifeforms, prokaryotes, to start in the ocean - another 2.08 galactic orbits - arriving at the date of 5.00 GY. Another 3.75 galactic orbits passed – 900 million years – before the development of cyanobacteria (formerly known as blue-green algae) started emitting oxygen into the Earth's atmosphere at 8.75 GY, 2.5 billion years ago. This oxygen release was done over another 2.08 GY through photosynthesis one invisible atom at a time from invisible one-celled organisms. At 10.83 GY (2 billion BP) the atmosphere had finally been infused with 21 percent oxygen, enough to cause fire combustion, and it has more or less stayed at that oxygen level since. Earth already had more one-celled organisms than there are individual stars calculated to be in the known Universe and was 57 percent its current age.

The kingdoms of life remained in the ocean for about 3 billion years until algae and fungi spread onto land 450 million years ago, at 17.29 GY, allowing terrestrial plants to evolve. The Age of Dinosaurs began 220 million years later and lasted a long span of about two-thirds of one galactic orbit, from 18.22-18.90 GY. Only 27 percent of one galactic orbit has passed since the dinosaurs' demise 65 million years ago; in fact, one GY has not yet completed since their beginning.

The first humanoid, *Homo habilus*, evolved into existence over 2 million years ago, eight-tenths of one percent of one galactic orbit. More so, modern humans started about 50,000 years ago and developed agriculture circa 10,000 BP, after the last Ice Age, or about 2.4 ten-thousandths of *one* of Earth's 19.17 GYs. Our solar system essentially hasn't even moved relative to its orbit around our galaxy since we began to domesticate our food sources. More than our ability to use fire, develop language, or pass our knowledge on with the written word, agriculture is what has allowed our species to thrive since about 19.1658 GY, the dominant culture's birth.

When taking photos of Earth, Voyager 1 showed that Earth is almost an imperceptible pinpoint of light from as near as Saturn, which is relatively close astronomically, being only 800 million miles away at their closest orbital positions. Mercury, though faint, is naked-eye visible from Earth, but Earth is most likely invisible with the naked eye from Saturn. The distance to the center of our galaxy is 22.5 billion times greater than the distance from Earth to Saturn. Buckminster Fuller referred to Earth as a spaceship; rather, it is an oasis teeming with life in a pristine interstellar desert, a campground in the cosmos and, relative to the Universe, is utterly invisible.

And yet life has caused a way, in human form, to be consciously aware of Creation. In other words, though we are invisible, relative to the Universe, we have the importance of giving realization to the Universe. Our next extraordinary achievement, in 2018, is to launch the James Webb Space Telescope into orbit four times further out than our moon is with technology that will be able to see further and better than even the Hubble Space Telescope has; another 4 percent of deeper sight may allow us to see back to the Big Bang itself.[10] Certainly, it will expand our understanding of our astonishing Universe and whatever else doing so leads to.

Back to Earth, an event is happening at all times as our planet spins on its axis. This constant event is called the terminator, the transitional period between night and day during dawn or dusk, the interplay of light and shadow caused by the Sun as the Earth turns, and terminating each day and night. At all times around the world the terminator sweeps over the Earth, a continual wave with each revolution of our planet, a planet which never rests. Ever-greater numbers of people continually arise at dawn everywhere to begin their daily activities, both legal and illegal, cutting trees, building, trafficking wildlife, clearing and burning forests, spraying chemicals, and an excess of millions of other similar undertakings of overconsumption combined with overpopulation. Forgotten in the daily lives of each individual is the ubiquitous, escalating event occurring at all times: the diminishment of the Earth's resources much more than it can now return, and worse, the demise of its ineffably precious species.

DOMINANT CULTURAL IDEOLOGY
(DCI, DC, DIC, DI)

T HE DCI BELIEVES THAT humans are the reason the Earth was created, that the Earth was created for man. With little erosion, and in spite of uncountable discoveries, including evolution, the DC still believes that man was created specially by God, separate from all other creatures, and that humans are the final apex of all creation, a finished product created in God's image.

The DC believes creation was complete once humans appeared. These mythic beliefs began thousands of years ago and have yet persisted: humans once believed the stars and the Sun revolved around the Earth, as infants believe everything revolves around them, yet members of the DIC still believe the Earth exists solely for their benefit and that nothing else matters.

Being made in the image of God, the dominant ideology (DI) also thinks God is some kind of human, albeit a superhuman, and that God is our potential. This means there's no other life in all of creation greater than us, certainly not on Earth. Thus all other life on Earth is far inferior; as a result, the DCI sees all life on Earth as being made for the purpose of serving humans, and any species that don't serve humans can therefore be eradicated. Amazingly, even the extinction of such species is deemed as inconsequential or at least acceptable.

Although the DC believes we are made in the image of God, it also believes we are flawed, that we're born impure and in shame and "sin" (since we had to come through an act of sex like the rest of the animal kingdom we're supposed to be separate from). Our flawed but ever-powerful human nature is such that we are detrimental not only to ourselves but to all life - but it's ok because "God's allowing it," so like adolescents, the DI still doesn't have to worry about being responsible, and a parental God is still in charge.

Our DCI paradigm insists we were created in God's image but have a flawed human nature, yet this is backwards. We inherited our human nature from the natural world, stemming 3.4 billion years back from the roots of our DNA, which arise from the invisible Kingdom of Monera (bacteria). God thus created our nature, but we have created our own image from our own hubris and our extremely limited, immature, and archaic beliefs of what God is. In our self-importance, apathy, and lack of empathy, we have taken full advantage of all other life on Earth.

What separated us from the rest of life was agriculture, and it is from agriculture that the DC was born. Through agriculture, we attained the power to control our own food supply and create surplus, a feat no other species has accomplished, and we steadily freed ourselves from being at the mercy of the seasonal whims of nature, formerly believed to be controlled by "the gods." The DIC thus gained reasonable control and power over nature, which it believes it's here to subject, improve upon, and put in order (though nature was already in perfect order). Such power to control our food, and thus our survival, allowed the DC to believe it owned the world; it is through agriculture that we believed we had dominion over the Earth and all its biota and made it possible to "be fruitful and multiply."

With agriculture came the need to protect cultivated food sources, and this we continue to do with earnestness. The DI operates within a paradigm of conquering - conquering land, people, nations, everything. Conquering land meant replacing wild species with domestic species, both animal and plant. The DCI's ranchers and farmers for centuries have systematically killed off everything we can't eat or anything it thinks interferes with what we eat. This is done as if it's God's will, and U.S. government services greatly assist them to do so for free through the USDA's Wildlife Services Program. The more competitors destroyed to protect the food supply, the more humans that can be brought into the world. The DC's mission is to support as many people as possible on Earth because we are the favorite of God and the reason "He" created the Earth. As a result, biodiversity is progressively diminished in order to support the expansion of one species: ours. "If there is one notion that virtually every successful politician on earth – socialist or fascist or capitalist – agrees on, is that 'economic growth' is good, necessary, and the proper end of organized human activity"[1]; growth for growth's sake is so well established that there's no tolerance to even question it.

With the need to protect our food sources came the idea that property could be owned, and the concept of absolute property rights arose when it was realized the route to personal wealth was land ownership. This ideology has led to corporate land ownership, the second-most destructive human device ever conceived. The concept of absolute property rights has proven itself to be even more destructive than the atomic bomb and is an obsolete philosophy continuing to lead us into the mass extinction of species. Since the environments of Earth will outlast every generation, for anyone - especially "persons" of enormous resource extraction such as multi-national entities (MNEs) – to have the absolute "right" to turn vast areas of environment to ruin, from having claimed and been given "ownership" to do so, would be highly questionable in a sane, self-preserving culture. Such ownership is no more than a manmade concept; along with economic theory that insists on incessant growth with little regard for the future, it is causing a race to complete environmental devastation. Incessant growth is especially poignant in agricultural economics because demand only grows with population, since there's only so much food a person can consume, unlike other consumables (products).

As long as it believes the world was made for humans, the DC will never admit that human overpopulation is a problem; even if it does, it's God's problem, not ours. The DIC believes it is the right of individual humans to bring in as many children as they want; room just needs to be made for them by simply moving all other life out of the way, since other life is of much less importance anyway. It's all about us after all, and to the DI, having as many babies as possible is doing God's work, which is a great excuse for being irresponsible with the overpowering nature and pleasure of sex, because at least anybody can then do "God's work," even more so if some religion tells them to. The DCI is still in massive denial that human overpopulation is the number-one reason why the ongoing mass extinction of species is happening; moreover, the

DC doesn't care because they don't yet see how it applies to them. As long as the prosperous can take and have as much as they want, they see no real problem. And the destitute are left to think it's God's will.

With agriculture came the patriarchy; previously, cultures were matriarchal, as it was women who were seen as the creators of life and were therefore revered and held the power, but with agriculture, matriarchal power was usurped. The DIC is a patriarchal culture, wrested from the matriarchy in the Middle East where agriculture began, setting the foundation blocks for the growth of the DI. Patriarchal control began 10,000 years ago with the onset of agriculture, and it established itself nearly 6,000 years ago with Judaism, which is why the Abrahamic religions say the world is only 6,000 years old. "Official" dominance came about 3,500 years ago when the patriarchs produced the first religious writings still used today. (The story of Genesis is included in the Torah, Bible, and Qur'an). Religious or not, the DCI believes human history didn't begin until the onset of agriculture; anything before that is nebulous and unimportant anyway, for all was Stone Age until the DC learned "the right way to live."

Once the DIC believed it had found the answer on how to live and had the power to enforce those beliefs upon others, the DI started taking over the world with the ability, through agriculture, to feed armies and to support more people. The DCI evolved into "might makes right," or "our way is right because you can't stop us," or "submit to our way or die," or "we have a right to your land because we are the only ones who know the right way to live," and "our God favors us because we always win (even if we have to abandon our own sacred edicts to do so)." The DC has been intolerantly telling all other cultures how to live since its inception, yet it has taken the entire world into the sixth mass extinction on Earth.

Cultures around the world support an extremely patriarchal paradigm to the point of severe male chauvinism. This chauvinism (which includes women since they are just as immersed in the beliefs of the culture as men), sees things as either/or; there's not a lot of room for the in between. It has long been, "my way or the highway," and "only the strongest survive," and "women are lesser than men and are here to serve and obey men," and worst, "women are the root of all evil" (from the Adam and Eve story). Furthermore, this patriarchal culture says all of nature is subservient to men and is also here to serve as chattel, as property which men can do with as they please. Men have been the decision makers, providers, and leaders (by force) for at least 6,000 years. How many of the hundreds of thousands or more of popes, priests, rabbis, imams, dalai lamas, and other religious leaders that tell everyone what to do, how to be, and how the world is meant to be have ever been female?

In the early 1600s famed philosopher, Rene Descartes, declared that animals were "nothing but machines, incapable of feeling pain" as part of the "Cartesian logic of the Enlightenment." Later that century, Isaac Newton proposed his cause-and-effect theories of reality. Since these two, the DIC has seen the Earth and its lifeforms as "a system resembling a machine."[2] Before that, the Church had long taught that nature was Satanic, full of pain and gore and "red in tooth and claw," and that life had been free of all of that in an imaginary Eden, the utopian fantasy where animals and humans got along in docile innocence and purity. The DI came to the Americas from Europe with hatred and terror of the wilderness and went on a massive killing spree; they wanted nothing more than to wipe out the vicious, terrifying, Satanic wild and make life safe, orderly, and more of what they thought was Godly and Eden-like. The DCI's objective has always been "to make human societies as independent as possible from the natural world and to make the natural world as subservient as possible to human decisions. Nothing was to be left in its natural state."[3]

Once the world was seen as a mechanism, materialism and consumerism became the law of the land, eventually enforced by big business. Wealth and fame became the new aspirations; corporate leaders the new priests, forcing their desires into government; and pop-culture stars, run by shadow organizations, the new social role models. The DC replaced morality for the pursuit of money, completely losing its heart for the spirit of and connection to the natural world. The DIC thinks nothing is of value unless it's for humans. It believes the Earth is a piece of property made for humans, nature is our nemesis, and certainly that protecting or preserving it is a hardship, and every tree, every rock, every fish – everything - has a price tag on it. The natural world then changed from one of spiritual wonder to one diminished to vapid commodities, to "economic units." "As long as the Earth is viewed as the personal property of the human race, a belief embraced by everyone from born-again Christians to Marxists to free-market economists, we are destined soon to inhabit a biological wasteland."[4]

The approximate six-thousand-year-old, patriarchal narrow-mindedness runs deep and is still very prevalent in our world, infusing our collective and personal unconscious. It still dominates the world: men have written, rewritten, and interpreted the bible many times – and even then misunderstand one of the fundamental tenets in Genesis with the now-infamous line of "be fruitful and multiply and have dominion over the Earth." Dominion does not mean domination; it doesn't even mean domestication. It means you have the power to rule, so take care of life in your kingdom, and be a cooperative, understanding, wise, responsible steward. Dominion works with the forces of nature; domination works against them. Dominion is symbiotic; domination is antagonistic. Dominion seeks to cooperate and innovate; domination seeks to control and manipulate. Dominion is compatible; domination is combatable. Dominion causes order and balance; domination causes chaos and imbalance.

The Native Americans, for instance, as well as any other indigenous people, lived in dominion more than any other culture on Earth. However, the arrogant, dominating, fruitful culture judged them as sub-human, so it was ok to treat them with disdain, disregard, pain, and death and learn nothing from them. The DI has treated the entire Earth this way, and the DC paradigm will soon result in a level of death unheard of in the Earth's 4.6-billion-year history: *the most immediate mass extinction ever known*... unless we wake up *now*.

If this world is to avoid the worst of the HOME, the Human-caused Ongoing Mass Extinction, we are far overdue in the need to upgrade our religious beliefs and dominant cultural ideology, and the Earth needs human unification for this effort to succeed. We are a species of amazing collective intelligence who continue to unravel the mysteries of life every day. We're also unraveling the tapestry of life. The DC is not being cooperative with the rest of life and has begun a long-term cycle of destructive competition and chaos.

However, the very controlling DIC mindset is not going to change on its own even when overwhelming evidence shows it's so very flawed and devastatingly wrong. The patriarchy will never admit they're wrong even unto their own or the world's death. You're either right or you're wrong – and they're always conveniently right. In the extreme, they'll even "prove" they're right by killing you. The current paradigm has caused an immense amount of pain in the world to both humans and non-human species.

The DI world divides life into superior or inferior and human or sub-human, so either some have rights or they don't, and some either matter or they don't. DIC pain matters but others' doesn't, whether it's human or non-human. Incredulously, even today, DC scientists are still testing animals to determine whether they can feel pain, torturing them in the process. This, in spite of the fact that other scientists have proven decades ago that even one-celled organisms have receptors for and create their own endorphins. The over-rationality of certain humans that kills emotion and empathy is one of the symptoms of the patriarchal paradigm we live in. These kinds of men and women are able to skin animals alive for their fur (to make Angora sweaters, for example, one can get five furs from one rabbit, if the rabbit lives each time, rather than if it's killed for one fur), hack off the faces of elephants they've machine-gunned, bludgeon screaming dolphins to death, and thousands of other atrocities. This is how the DIC paradigm treats our entire Earth, as if it's just an unfeeling mechanism subservient to man, still adhering to the fallacious Cartesian philosophy expounded 400 years ago. Will the DI ever admit it has been wrong?

More likely, even in the face of unmitigated catastrophe, it'll say, "Let's wait and see; it ain't over yet." Since Genesis, the DCI has wrongfully believed it's exempt from the laws of nature because of its belief in being specially created by and given special status from God. At this point, these beliefs in exemption are forms of massive denial. One form is secular and humanist, which says we don't have to change because human genius will provide; the other is religious, which says we don't have to change because we are in the hands of god.[5] "We have fallen prey to the illusion that we can modify and control our environment, that human ingenuity insures the inevitability of human progress, and that our secular god of science will save us,"[6] which is humanism.

Humanism is the secular outgrowth of DC religion but still believes the world has been made solely for human benefit. It's also the modern deification of Cartesian and Newtonian philosophy whereby the Universe and life function essentially as a machine. Humanistic ideology is a sect of the DIC based on the use of reason, the scientific method, and the denunciation of spiritualism. It believes human intelligence is so capable that we can fix all of our problems on our own and assumes an all-seeing God is non-existent. Humanists anticipate the success of artificial intelligence (AI) will ultimately replace any and all past notions of God. However, humanism's blind spot is that, although it claims we are genius, its lofty mentality arrogantly fails to value the function of nature and our relationship to it. Extinct species are unimportant because humanism believes it can replicate the functions of those species when and if needed, such as the creation of robotic bees to extend its toxic ways of agriculture, rather than change.

Many humanists dream of the creation of a fully humanized, artificial world as the glorious fulfillment of human destiny. Humanism is the ultimate gratification of the ego, and egos of the DI are destructive. They'll even walk on the furs of endangered animals used as *rugs,* yet panic if a bee flies near them. They are not at peace with nature whatsoever. Humanism operates only on logic and practicality, so protecting nature must have such reasons; it sees nature in reductionist terms and as an economic machine, so the very idea of protecting it for any non-human use is oxymoronic.[7] With humanism, the needs of humanity will always supersede the protection of nature, though it does not see that protecting nature is our highest need. As example, the humanists carelessly believe in the very implausible idea that we will be able to reconstitute some extinct species from vials of preserved DNA whenever we want and reintroduce them into the wild - a wild which already doesn't exist, which is why 90 percent of the extinct in the past sixty years or more are already gone.

Another example: the idea of decreasing the human population is unthinkable and completely contrary to the ideology of the DCI since the book of Genesis, even though it is essential to the protection of nature and our own well-being. Such an idea is blasphemous, and conjures up fears of holocausts conducted by a cartel of monstrous tyrants. No one ever imagines even the possibility that we could make a long-term, conscious, intelligent choice to gradually bring our population to a level of true sustainability without barbarism. Once again, it's either/or: either unrestricted freedom to have any number of children or unacceptable control of such "rights." Reevaluation of our entire paradigm is crucial to the well-being of our Earth, yet it remains unthinkable and untouchable.

Relying on science or politics to reevaluate our paradigm will also take too long, if it were ever to even happen, because science and politics are opposed to each other. Science seeks the truth, while politicians, in order to keep their jobs, gratify corporate special interests, which are opposed to the truth in regards to the environment because such truth is bad for business. Furthermore, environmental debate cannot be resolved by science because "[t]he whole direction of science since its birth in the Renaissance has been to master and exploit nature for human benefit."[8] Plus, science may be able to point out what's happening, but whether it's perceived as a problem depends upon one's values[9], and the DC only values nature economically. Decisions are made based upon our values, which are based upon the myths we live by, the ideology. The myth of our misunderstood dominion isn't working; rather, it is leading us to various forms of depravity.

Charles Darwin cracked the DCI with his work that led to the proof of evolution and to the development of humanism, but it also led to misleading slogans such as "survival of the fittest" and "greed is good," which are somewhat softened by the names of "social Darwinism" and "corporate Darwinism." Survival of the fittest only acknowledges competition and completely ignores cooperation, the higher truth of a more functional, sustainable, and compassionate world. By twisting Darwinism, the DI mindset told itself it was ok to exploit others, that it was perhaps even speciating from the rest of humanity. Social Darwinists then gravitated to the MNEs where they could flourish within the belief in corporate Darwinism, the most powerful aspect of the DCI and thus the latest crown jewel of DC ideology, where elitism or self-defined supremacy, arrogance, and rationalized lack of compassion run rampant and are idealized at the pain and expense of all other life. Such Darwinism may soon be upgraded to transhumanism: technologically enhanced humans.

The DIC thinks "virgin land... *needs* to be taken over and cultivated; it *needs* to be made productive for humankind. There is something immoral about leaving it alone, letting it lie there untouched, un*improved*."[10] Just a few famous quotes that represent the DI: "It is in the power of future technology, perhaps in accord with divine providence, for people to flourish as never before in a completely humanized environment, a paradise of our own making. Such is the foreordained trajectory of an advanced intelligent species. I tell you, it is our destiny!"[11] "The destiny of man is to possess the whole Earth; the destiny of the Earth is to be subject to man. There can be no full conquest of the Earth, and no real satisfaction to humanity, if large portions of the Earth remain beyond his highest control." - John Widtsoe, a Mormon patriarch.[12] "Man is accustomed... to consider himself the crowning creation of Nature, why should he not believe that he represents also her final purpose?" – Goethe.[13] "[F]orestry experts saying, 'The trees are ours for the taking and damn, they must be taken lest a resource go to waste.'"[14] "The Gulf of Mexico is a very big ocean. The amount of oil and dispersant we are putting into it is tiny in relation to the total water volume," - Tony Hayward, British Petroleum's CEO during the catastrophe of millions of gallons of oil billowing out of the ocean floor for thousands of continuous hours into the Gulf of Mexico.[15] "Man, if we look for final causes, may be regarded as the center of the world, insomuch that if man were taken away from the world, the rest would seem to be all astray, without aim or purpose," and, "I am come in very truth leading to you Nature with all her children to bind her to your service and make her your slave" - Sir Francis Bacon.[16] And the most obvious oxymoron, "The first great fact about conservation is that it stands for development" - Gifford Pinchot, founder of the United States Forest Service.[17] Such quotes are endless within the dominant cultural ideology.

EVOLUTION

T HOUGH THE FIRST MAKINGS of DNA may have originated from beyond the Earth, even if we ever find extraterrestrial life, our Earth surely has its own unique evolutionary expression unlike anywhere in the Universe.

After the 700-million-year Hadean eon of fire, when the Earth was forming as a molten orb, it slowly cooled enough for water vapor to condense and fill the oceans. The oceans remained lifeless for another half billion years, as atoms mixed in an aqueous, anaerobic world until a successful molecular combination formed into a ribonucleic acid (RNA) molecule. As RNA molecules have proven to evolve in test tubes, it shows that biochemical evolution prospectively led to the formation of the deoxyribonucleic acid (DNA) molecule and the beginning of our Earth's life expression. How this miraculous phenomenon exactly occurred remains a mystery, but we do know that DNA took its first forms of life as single-celled prokaryotes, as aquatic bacteria. RNA since took on the role as an intracellular messenger replicating segments of the DNA molecule, and DNA is the message, scripting every organism's existence.

The DNA molecule was not discovered until 1953. Fifty years later, the extraordinary technological achievement of fully mapping out the human genome was completed, and in 2007, Craig Venter, who led the successful project, made history when he became the first individual human to be fully decoded, bared to the world. The human DNA molecule is only twenty hydrogen atoms wide but when unwound is over three feet long; it records 3 billion base pairs of information, which is more than thirty complete sets of the Encyclopedia Britannica. In each cell.

In the process of genome research, many other organisms have been decoded - bacteria, yeast, worms, mice, chimpanzees, sharks, cows, turkeys, sea urchins, rice, wheat, corn, etc. – which has revealed that all lifeforms on Earth are related, at minimum, by the same fundamental genetic sequence, from the very first prokaryotes to the most recent, including humans. Of our approximate 25,000 genes, we share a majority with mice, about half with fruit flies and roundworms, and more than 1,000 with yeasts. We share a few hundred with bacteria, and this foundation of genes is universal to all living organisms; they encode the "…basic life functions [of] DNA replication, the production of proteins from RNA, energy metabolism, and the synthesis of… adenosine triphosphate, or ATP, the energy currency for all life on this planet."[1] We have proven we are genetically related to every other organism on Earth. Within us we literally share genetics with every other form of life; we are intrinsically part bacteria, part yeast, part Ponderosa, plankton, blue whale, chipmunk, goldfinch, giraffe, grass, grizzly bear, sturgeon, mushroom, starflower, starfish, sea urchin, tiger, and all the millions more that exist. We share the full history of life on Earth in our genes. The history of life on Earth is innately who we are and is the basis from which we have evolved in every minute detail. We are *profoundly* the exact opposite of being separate from nature, as the DC believes, or the Earth and all its dynamics.

With the development of the DNA molecule, life began to slowly build upon its successes of invisible atomic structuring, each step, each challenge overcome, recorded in the chromosomes of each cell within every lifeform. Life force in the form of DNA has stimulated the creation of millions of fantastic evolutionary changes throughout the 3.4 billion years of life on Earth. In its undying mission to ensure its own survival, life, through DNA, found the means to explore unknown stimuli, encouraging evolution, which formed all species' senses and specialized forms.

Before sexual replication, organisms asexually cloned themselves; without genetic variation this made them easy targets for large-scale parasitical attack. Causing genetic variation, sex retarded parasites' capabilities, allowing hosts to better survive. With time, however, successful parasites - ones that didn't kill their hosts - evolved into becoming sections of the host's DNA. As a result, about 40 percent of our human DNA shows the history of this ongoing, metamorphic exchange.[2] In other words, parasites are one of the primary causes of evolution.

Genetic evolution doesn't just add to what already exists, it also takes traits away from organisms for purposes of improvement or speciation. In a process called apoptosis, certain larvae turn into butterflies through structured tissue dying off. The human fetus likewise passes through phases of apoptosis in the womb, proving our evolutionary roots. In the case of speciation, entire genetic sequences may be added or subtracted over time.

The Kingdom of Monera (prokaryotic bacteria), the first kingdom of life and of non-nucleated single-cell organisms, existed for about 2.2 billion years (a long 9.17 GY) before multicellular life began. As the trunk of the tree of life, this bacterial kingdom is the foundation from which the other four kingdoms of life branch off (speciate). Prokaryotes led to and are integral parts of the Kingdom of Protoctista (nucleated single-cell microorganisms), the Kingdom of Animalia (from zooplankton to higher vertebrates), the Kingdom of Fungi, and the Kingdom of Plantae. Bacteria are part of these other kingdoms as much as xylem and phloem are to a tree. Each kingdom branches off into phyla, classes, orders, families, genera, and species with each progressively descriptive category increasing in numbers.

The bacterial Kingdom of Monera and the eukaryotic (nucleated) microorganisms of the Kingdom of Protoctista are the processors of chemicals at the level of atoms and molecules, the job of fungi is absorption, the role of plants is production, and the function of animals is consumption; all cycle back to the monera and protoctists.

As the sole kingdom of life for eons, many forms of bacteria evolved with one another, consuming each other and merging their non-nucleated, undigested genes. Some bacteria, such as mitochondria, became parts of other bacteria. A mitochondrion is the part of each cell of life that processes and provides energy to the cell such as the oxygen in our own human brain cells. "The oxygen we breathe enters the brain from our bloodstream and is incessantly metabolized by the mitochondria that we know are former respiring bacteria."[3]

Mitochondria have still retained their own separate DNA code that proves it was once its own individual form of bacteria. It even replicates within and independent of the cell in which it exists, as do other plastids, such as chloroplasts, which perform photosynthesis. Crucial bacterial mergers or integrations that exchanged genes and created new types of complex cells happened through a process known as symbiogenesis. Such symbiogenesis led to the development of protoctists.

Four particular and fully distinct types of bacterium partially ingested one another until successful processes of life eventually formed from the merging of their DNA. First archaebacterium fused with swimming bacteria, creating nucleocytoplasm. Later, this combined with mitochondria/oxygen-breathing bacterium. The fourth component to be partially ingested and integrated was photosynthetic bacteria. Partly devoured bacteria trapped inside the bodies of other bacteria became organelles of new cellular lifeforms, which further developed into the ancestors of plants (chlorophytes), animals (zoomastigotes), and fungi (chytrids). Again, the kingdoms of animals, plants, and fungi all evolved from protoctists which originated from the successfully combined biogenetics of prokaryotes. "All nucleated organisms... arose by symbiogenesis... [which] is now indisputable"; the late Lynn Margulis, the brilliant and pioneering scientist and first wife of Carl Sagan, explained this well in her book, *Symbiotic Planet: A New View of Evolution*.[4]

For eukaryotic, multicellular lifeforms to appear, prokaryotes had to be able to metabolize energy from the Sun, carbon from the atmosphere, and hydrogen from the sea, a long process that took up nearly half the age of our Earth after the Hadean eon and the half billion years of lifeless ocean. Multicellular life didn't begin until 14.17 GY, 3.4 billion years since our Earth's beginning. Until then, life was invisible in the ocean and all land was utterly barren. It took another 2.8 GY (670 million years) for life to develop skeletal forms and shells, which is one of the defining traits of the Cambrian Explosion (of life) around 540 million BP. For that to happen, cyanobacteria, the first photosynthesizers, had been releasing atoms of oxygen into the atmosphere for at least 1.3 billion years, which also created the protective ozone layer, after having phased out the dominance of anaerobic archaebacteria. To this day, prokaryotes are still more tolerant of ultraviolet radiation than eukaryotes, since they had survived without ozone protection for eons, protected from UV radiation to a lesser degree by the filtration of the seas. UV radiation damages DNA, so until about 450 million BP (17.29 GY), life was limited to aquatic environments until there was enough ozone in the stratosphere.

Life existed in the ocean for almost 3 billion years before moving onto land. The Kingdom of Fungi spread onto land interconnected with algae, spreading mycelium like a nervous system throughout the soil, known as hyphae. As absorbers of nutrients, fungi were the catalyst that allowed plants to take hold on land. Fungi and plants have had a symbiotic relationship since the beginning of their terrestrial presence. The oldest fossils of fungi are intimately connected with plant tissue.[5] "[G]reen land plants descended from green algae... they were rootless, leafless, and stemless but upright seaweedlike organisms."[6] Today, mycorrhizal fungi are attached to the roots of 95 percent of the world's plants, spread under the soil in a network of mycelium. Mycelium is what the "fruit" of the fungus arises from, such as mushrooms, and can grow to networks of thousands of acres large, considered a single organism. After their colonization of land, spore-bearing mycorrhizal fungi were instrumental in giving rise to spore-bearing ferns within about 50 million years. Over a period of close to 1 GY (240 million years), it is the development of seed ferns (and insects) that eventually gave rise to angiosperms (flowering plants) circa 140 million BP.

Life is always in pursuit of food. With vegetation on land, DNA took about 70 million years to figure out how to get to it with the establishment of amphibians by 350 million BP. It was first in fish that life evolved jawbones about 410 million BP. Arthropods, which developed during the Cambrian Explosion, also speciated out of the sea to inhabit land in the form of centipedes, springtails, and mites, among others. Some millipedes and such grew to be several feet long, as there was almost 50 percent more oxygen in the air that allowed them to grow so big.[7] Reptiles speciated out of amphibians during the time of the Carboniferous period (so named because of the creation of coal deposits from the abundance of fallen lycopod trees and such) 350-275 million BP, and they expanded their diet to include bugs.

During the Permian period (circa 250 million BP), life had just gotten fully established on land just in time for the biggest mass extinction on record. In recovering from that, more complicated arthropods emerged such as lobsters and the first insects. Modern coral also took hold.

By the time of the last mass extinction around 65 million BP, mammals and the other classes of life that exist today were established. The asteroid impact eventually destroyed enough food resources on a global scale that larger animals could not maintain themselves, plus they had too long of reproductive cycles and thus perished.

During the age of dinosaurs, mammals came into being, existing small and nocturnally. Mammals have the unique quality of a four-chambered heart with complete double circulation, which disallows aerated blood in arteries mixing with oxygen-depleted blood in veins, a remarkable evolutionary development.[8] Mammals are known as homeotherms, meaning they require more food calories, as heat is generated internally by metabolism with more mitochondria. Heat is further regulated with hair growth and sweating. However, at the time of the event that finished the dinosaurs, it is their small size and faster procreation rates that also allowed mammals to survive.

With the onset of terrestrial angiosperm plants, plants and animals evolved to have more similarities to one another than any other lifeforms on Earth; they are the only ones that develop out of embryos, both produce eggs and sperm (pollen's equivalent), both have vascular systems, both communicate chemically to the inner and outer world, and both emit electrical impulses for intercellular communication (though without nerve cells, this function is rudimentary in plants). Physical nerve cells evolved much later and only exist in the Kingdom of Animalia, so animals respond to their environment both biochemically and bio-electrically.

The first communication system within a cell was chemical, consisting of simple molecules now known as peptides. Even one-celled organisms, such as the tetrahymena (the "workhorse of biology"), has peptides common to humans such as insulin and beta-endorphins (pain killers) and their matching opiate receptors.[9] Peptides are such chemicals as adrenalin, hormones, and oxytocin, the chemical in mammals that causes the contractions for birth and of bonding through physical intimacy throughout a healthy life. Peptides comprise the primary intra- and intercellular communication system of life and exist within all organisms. There are hundreds of known peptides that form within humans, including those that are part of our emotional nature. Every cell in our bodies both release and form receptors for peptides, and all cells are responding and reacting to one another in this manner.

Science has recently discovered certain lifeforms whose systems function solely on these ancient chemical-peptide substances and have found very powerful medicinal uses for them in humans. Unfortunately, as in the case of cone snails, many such species are in grave danger; for instance, cone snails live in tropical mangrove environments, which are being destroyed as rapidly and in like manner as coral reefs.[10]

Of urgent concern is that our human population and our economic and cultural ideologies and activities are putting an end to the process of evolution; furthermore, we are increasingly erasing much of the genetic record of life on Earth permanently and deciding which species live or die, as most still believe we are separate to the rest of life or even superior to it. Our Earth is forever changed by us, and our genetic setback to life is becoming ever more severe.

OVERPOPULATION

POPULATION IS A CONCERN of numbers. The global population of people reached one billion for the first time in Earth's history only 216 years ago when the average human life span was about thirty-seven years old. There were about one million humans by the time agriculture began at 10,000 BP. It then reached 250 million by 2,000 BP. After fifteen years of bubonic plague killed a fourth of Europe's 100 million people, ending at 654 BP (1361 AD), the world's population was around 350 million people. It then took 439 years for no less than eighteen generations to add 650 million more by 1800, our first billion. Another billion was added in just 130 years (1930), another billion in thirty years (1960), the fourth billion in fifteen years (1975), and the last three billion at basically twelve-year intervals (1987, 1999, and 2012). We are now creating one billion humans 5 million times faster than it took the first billion to exist (a 12 to 2.5-million-year ratio). Even at an expected slower birth rate of 1.6 percent (2.1 formerly maintained the population), if nothing changes, it will only take about ten years for the next increases of a billion: 8 billion by 2020, 9 billion by 2030, 10 billion by 2040, 11 billion by 2050.... The best-case scenario predicted by the UN in 1998 was 7.4 billion people by 2050; we've already reached that in 2015. The UN's worst-case estimate was 10.7 billion by 2050.[1]

Since the signs are already here, we could all realize our population growth is unsustainable but for the DC's mass denial, which uses tools such as nonstop disagreement and shame to marginalize intelligent problem-solvers to push unfavorable, unpopular action further into the future. Combined with religious dogma, the DC collectively throws common sense out the window concerning overpopulation; or others believe the humanists will once again save the day; and still others think the population bomb is a non-issue because the timing of all the predictions of doom from the 1970's never happened, supposedly proving those speculations wrong. Whether we ever reach 9 billion and beyond, it's likely we'll try our best to do so, since the DI believes in "more people no matter the consequences." It is also certain that the mass extinction of species will continue until we change our way of thinking, as no authorities will have the mandate to seriously address our number-one fundamental problem: overpopulation. Instead, we'll let nature run its course the way all other species have done – all those unintelligent species the DCI believes it is so above and separate from. "If current predictions of population growth prove accurate, and patterns of human activity on the planet remain unchanged, science and technology may not be able to prevent either irreversible degradation of the environment or continued poverty for much of the world." - The Royal Society of London and US National Academy of Sciences.[2]

We are not using our intelligence on the issue of overpopulation because the DI disallows it, subtly or not so subtly influenced primarily by religion as the historic and collective moral compass that still has control over modern mass consciousness. The DCI is the socially acceptable default mechanism that gives simple, undemanding answers to our overpopulation predicament: we simply "turn it over to God" so we don't have to deal with it, thus abandoning our responsibility, enabling ourselves to keep doing what we've always done, which extends our destructive DIC to the point of crisis and extinction. Religion, as the long-standing foundation of the DI, insists that people have as many children as they can or want regardless of any consequences, and that the ensuing population must be provided for, again, regardless of any consequences, ultimately taking the form of catastrophic environmental destruction. The well-being of humans is always the top priority, and the rest of nature is ignored as unimportant in relation to that priority. This insistence and priority is creating the exact opposite of the well-being of humans; for instance, 3.4 billion people don't even have access to a toilet and live on less than two dollars a day. Even though the majority of children being born are not planned, expected, or wanted, people still having as many children as they do indicates how deeply denial, the normalcy bias, religion, and plain-old human selfishness or ignorance are entrenched, and it's undeniably presenting unacceptable manifestations.

Unlike any other species, our ability to create and control our acquisition of food is the reason we have thrived to the point of taking over the entire world with our numbers, extensively destroying the wild, becoming the determining factor of evolution, and causing a mass extinction of species. The biomass of humans and our 70 billion domestic animals now comprises 97 percent of vertebrate weight, leaving only 3 percent of vertebrate biomass in the wild; it used to be the opposite.

China holds about 1.4 billion people; India, though it is only one-third the size of China or the U.S., still squeezes in about 1.3 billion; and the U.S. contains one-third of a billion. All three are the most populated nations on Earth, and every ten years the population of the world now increases by three entire U.S. populations. What part don't people understand about us having an overpopulation problem?

Already, we are harvesting up to 50 percent more natural resources per year than the Earth can produce just to keep up with current human demand; it now takes eighteen months for Earth to produce the resources we use up in twelve. The "bank account" of nature is being depleted. That's why our human population already uses more fish than the seas can sustain, crashing fisheries; uses more wood than the forests can grow, shrinking forests; has less grass to feed our livestock (primarily beef) than is naturally available, which is why up to 40 percent of precious farmland and water usage is used to grow crops such as hay, alfalfa, and corn for them; uses more water from underground aquifers than can be sustainably pumped out; loses more farmland than can be newly created, which entails destroying more forests; is why deserts are expanding and glaciers are melting; and is why there is 35 percent less of a non-domestic animal inventory on the planet since 1850 (along with a 600 percent human increase). Human demand equates directly to our human population, making any kind of further development unsustainable, for even in supplying our current demands we are over-drawing nature's "principle" when sustainable would mean living only on nature's "interest," interest being what nature can provide every year without depletion. Obviously, the way we are living cannot continue indeterminately, yet the real crisis is our human-caused *extinction of species*.

To present another overpopulation consideration, Americans use fifty times more of Earth's natural resources than the people of Bangladesh, a country smaller than the size of Iowa, yet the population of Bangladesh is 164 million (July, 2013), half the number of the 320 million people in the USA. America's population then equals 16 *billion* Bangledeshis in its use of natural resources, in this regard making it by far the world's most overpopulated country[3]; inversely, Bangladesh's population of 164 million equals 6.4 million Americans' lifestyle. If you think the population explosion is going to slow down, 33 percent of Bangladeshis and 25 percent of Americans are under eighteen years old, and this is typical for the rest of the world. Plus, the rest of the world wants to improve the comfort of their lives by acquiring at least some of the U.S. lifestyle. Imagine Bangladeshis attaining even one-quarter of the American lifestyle; that would equal the effect of *4 billion* people living in Bangladesh – or imagine Iowa with 4 billion people rather than its current 3.1 million.

Demographic winter is the term used to describe the impact of our overpopulation, a term borrowed from "nuclear winter," used to describe the aftermath of a nuclear holocaust, and most biologists compare what we are doing to the same effect as the asteroid that hit the Earth and destroyed the dinosaurs' world. In just 116 years, our population has grown from circa 1.5 billion to over 7 billion, presently increasing by a net of 230,000 per day. The world's population has more than doubled since 1960 and more than tripled since 1930, just eighty-six years ago.

The "developing" countries are where 80 percent of humanity lives, and their populations have grown from 300 to 700 percent in only the past forty years, places such as Mexico and countries in South America, Asia, and Africa. As much as this is causing large-scale human suffering, the suffering is far worse for the world's animal and plant species, now disappearing in enormous numbers; they are victims of massive human exploitation, are losing their habitat, starving, and even dying of exposure due to anthropogenic climate change. The loss of biodiversity in turn further throws off the balance of Earth's operating systems, causing a positive feedback loop (accelerating the same effects), which adds to the Human Exponential Event.

Whatever started DNA is what started the evolution of life. DNA formed us and awakened intelligence within us, but human activities have stopped the evolutionary process. Species are disappearing rather than speciating, and the species that remain are only the ones we decide are worthwhile to keep - if we can. This has brought on the end of the wild because all species are being managed or mismanaged by us and are certainly impacted by our environmental ruinations. We have become the primary evolutionary force of nature - and a very detrimental force as a result of our very narrow-minded view of nature and the importance of life's interconnectedness.

The term "sustainable development" is an oxymoron, a simple attempt by business to morph human activity into another acceptable idea, but unless we want nature to do it for us, we must decrease development regardless of economic impact. We must change our economic system so that slowing further development will not cause a global depression and the human suffering historically associated with it. The betterment of humankind believed to come from economic growth, which depends upon endless development, must cease being our sole, top priority. We must change the present-day paradigm in order to prevent worsening the current mass extinction of species.

We also need to stop calling countries either developed or developing. They're already all developed. They've either developed astounding numbers of people and/or have developed an overconsumption of resources; the result is the same. Relative to the natural resources humans use, we are already surpassing Earth's biocapacity to support our species by the equivalent of almost 3.7 billion people too many. The question is: how much reserve is in the Earth's environmental "bank account"? No matter how long it can last, it is already unsustainable and a matter of time before we lose the Earth's ability to support everyone. Developed or developing, all are ruining the Earth; we all share that in common. Instead, truthfully, we could call a country an over-fertilized country (OFC) or an over-consumerized country (OCC). The Earth cannot sustain any more of either type of development, and both have actually not brought us the quality of life we seek, unless the OFCs consider being overcrowded and poor a success or the OCCs consider hedonism as such. In either case, both would have to regard humans' stupefying and reckless annihilation, the extinction of mass numbers of Earth's magnificent species, as a success.

PREVIOUS EXTINCTIONS

L IFE STARTED OUT ANAEROBICALLY. Very slowly, photosynthetic organisms developed that emitted oxygen. This caused a fundamental shift in life, as the anaerobic organisms went underground, adapted to become aerobic, or became extinct. This was the first mass extinction, an overlooked extinction between invisible, prokaryotic lifeforms. It is an oxygenated environment that allowed visible, multicellular lifeforms to become abundant on Earth, starting entire new kingdoms on Earth after 1.3 billion years of oxygen buildup, suddenly blossoming with the Cambrian Explosion between 540-590 million BP (16.85 GY).

The fossil record reveals five mass extinctions since the Cambrian Explosion. The first major extinction, the Ordovician, happened at 17.21 GY (circa 470 million BP), before life was on land, and wiped out 50 percent of biological families and even more genera. Sharks and the most basic of terrestrial plants began 60 million years after this event.

The Devonian extinction occurred about 115 million years later at 17.69 GY (around 355 million BP), losing some early types of fish, 30 percent of animal families, and about 70 percent of all species. Life was still predominantly in the ocean except for the earliest forms of terrestrial plants, fungi, and amphibians. The end of this extinction marked the beginning of animal life colonizing land.

The most titanic extinction in the history of our world, the Permian, happened at 18.13 GY (250 million BP), just over only one GY ago, and just after life had finally become fully established on land. The Permian extinction is known as The Great Dying: 14 percent of the *orders* of life perished, including eight orders of insects; 50 percent of animal families; 80 percent of genera; and 95 percent of marine species. This was caused by a hundred thousand years of massive volcanic activity in what is now Siberia, filling the atmosphere and oceans with sulphur. This extinction is commonly known for the end of trilobites, the cousins of horseshoe crabs.

The Triassic mass extinction followed only 55 million years later at 18.35 GY (195 million BP). This event ended 35 percent of animal families, many being reptiles and mollusks, though modern-day corals survived. Dinosaurs had been flourishing for 35 million years and still had 130 million more to go.

The well-known Cretaceous extinction occurred at 18.90 GY (65 million BP) with an asteroid that hit the southern part of what is now the Gulf of Mexico, coming in at a low angle from the southeast, sending a type of lahar all the way up into Canada. The event is estimated to have been equivalent to 70 million Hiroshima bombs. Well-recognized for the demise of the dinosaurs, any animal species over 55 pounds also went extinct.[1] Mammal species were small at the time and could thus forage enough food to endure the holocaust. This mass extinction took 75 percent of Earth's species. Sharks have survived four mass extinctions since they began as a species.

From Earth's four geologic eons encompassing its eras, periods, and epochs, all five mass extinctions occurred in the last eon, which began with the Cambrian Explosion, and all within just 1.69 galactic orbits. For the five kingdoms, from their phyla to their species, the biggest event in Earth's history is now: the HEE, our Holocene epoch, soon to be named the Anthropocene epoch, because our force of nature has the same impact upon species as an asteroid or of enormous volcanic activity. We're even killing off the sharks.

The two-million-year, interglacial Pleistocene epoch ended at 11,500 BP and began the Holocene epoch. The time of modern humankind began 50,000 years ago with The Great Leap Forward in our species' evolution. We've slowly but steadily caused the extinction rate to rise ever since, but once the European Golden Age of Discovery and the following Industrial Revolution took place, the extinction rates skyrocketed along with the ensuing and rampant exploitation of the natural world, which continues to escalate drastically. In our current wave of globalization, we are tragically finishing the Human-caused

Ongoing Mass Extinction of species. In sixty years or less, the Human Exponential Event will have run its course. "We will have turned Eden into Hell"[2] unless we wake up and change our ways immediately.

Earth was a molten ball of fire in the Hadean eon, named after Hades, the Greek god of hell. We could call our current epoch the beginning of the Neo-Hadean eon ruled by plutonium (named after Pluto, the Roman god of hell and death) and the age of corporate plutocracy.

In 2009, "3,246 of the world's animal species were classified as critically endangered."[3] By 2013, it had risen to 4,286, a jump of 32 percent in four years. That's 260 per year that are admitted, economically allowed, and non-centinelan. At least half of these will go extinct, which is a very minimum estimate that's 1,444 percent above the normal background rate *right now*. The "Goddess of the Yangtze," China's river dolphin, was declared extinct in 2006, the Western Black Rhino in 2013. The whooping crane is still critically endangered, and only 50 Javan rhinos exist, with zero in captivity. Eighty-three percent of endemic fish in the Philippine's Lake Lanao are extinct. Malaysia has lost 266 endemic fish species; Lake Victoria, fifty.[4] Since humans colonized the Pacific, around "2,000 species of Pacific island birds – about 15 percent of the world total – have gone extinct."[5] The list goes on and on and on for thousands and thousands of species, growing every day....

SPECIES TYPES

A SPECIES IS DEFINED BY the ability of those within its group to procreate fertile progeny, while those of different species may be able to mate and cause the birth of healthy lifeforms, but those offspring will be sterile.

Estimates of the number of species on Earth vary widely, but it's safe to say there are at least 9 million, not including unnamed prokaryotes (non-nucleated) and eukaryotic (nucleated) microorganisms. Science classifies about 10,000 more per year, mostly insects and oceanic creatures; most mammals and others have been found. The characteristics, habits, and ranges, etc. of only about 100,000 of all the discovered species are well-known.

Of the 1.6 million known species, the phylum of arthropods comprises two-thirds of all that currently live on Earth. Arthropods are invertebrates with segmented, jointed bodies and exoskeletons, such as lobster, arachnids, dragonflies, millipedes, ants and other insects, crab, and more. Arthropods started in the ocean during the Cambrian Period, evolving to later occupy the land and air. Phyla Arthropoda includes the class Insecta, a class that makes up 56 percent of all known living species. The second-largest invertebrate animal phyla of Earth are mollusks (clams, oysters, octopi, squid…) with over 100,000 species. Eighty percent of all the major groups of organisms are marine.[1] Half of all terrestrial animal species live in trees. Plants split from 300 families into 287,000 known species. The Kingdom of Fungi contains about 100,000 species, omnipresent and almost all terrestrial. The phylum Cordata (vertebrates) contains at least 54,000 species: approximately 29,000 fish, 9,000 birds, 5,000 reptiles, 5,400 mammals (including our one species of humans), and 6,000 amphibians.

Speciation is when one species develops out of another to take on another function in life, leaving the former species to carry out its already established role, such as when orangutans evolved out of gibbons.

Subspecies to a species are comparable to the different races of humans to *Homo sapiens* (though more so for non-human species). They're able to breed and produce fertile offspring but are exclusive due to geographical reasons or social/behavioral differences.

Centinelan Species are species we have not yet even discovered. There are untold numbers already extinct, with millions to go if we don't change our harmful ways. It wasn't until the 1980s that we even started discovering new species in the forest canopies of Northern California, and only after 75 percent of the world's original forests were already annihilated. There are untold undiscovered species in what tropical forests still remain, in the ocean, and surely in all other environments.

Keystone Species is a metaphorical term created by Robert Paine in 1966 at the University of Washington. A keystone is the central stone placed at the apex of a stone arch that supports the entire structure; if the keystone is removed, the entire structure collapses. Likewise, a keystone species has a disproportionate role of importance in an ecosystem that allows for thousands of other species to exist. Examples of keystone species are coral, for creating an entire underwater environment; prairie dogs, for providing a means for the restoration of groundwater, resulting in the health of rivers, and as a prolific food source for other species; beavers, for building dams that create new wetland environments for freshwater species; elephants, for creating the savanna by knocking down trees and greatly fertilizing and germinating the soil; and sea otters, for feeding on sea urchins, which would otherwise demolish kelp forests.

For the entire list of species types see the Appendix.

HOTSPOTS

IN 1988 THE ECOLOGIST, Norman Myers, coined the term "hotspots" to describe twenty-five precious areas on Earth that had the highest amounts of biodiversity left on the planet. Since increased to thirty-four, these formerly pristine, vast areas of wilderness are now fragmented islands of biodiversity that hold the most abundant of endemic species on Earth. To qualify as a hotspot, the area must have at least 1,500 endemic species of vascular plants and have already lost at least 70 percent of its original habitat (this number approaches an average of 90 percent in actuality). "[H]abitat usually means trees... when you take away the trees, a cascade of extinctions usually follows."[1] These irreplaceable regions contain more than 50 percent of our Earth's plant species and more than 35 percent of our Earth's terrestrial vertebrate species (mammals, birds, reptiles, and amphibians). However, these hotspots total only 2.3 percent of our Earth's land surface; alarmingly, only 11.6 percent of that small percentage is legally protected - and not very well.[2] Regardless of what happens anywhere else on Earth, if these hotspots are destroyed - a very real possibility - it means an astonishing amount of Earth's exquisite lifeforms will disappear forever. Already, even preservation is risky enough since there's nowhere for any of these species to go if future migration were needed due to any kind of environmental changes.

Of all the hotspots, Madagascar tops them all with almost 500 endemic *genera*. The Caribbean islands are next with almost 300 unique plant and vertebrate genera. The others are scattered around the planet, such as the Atlantic forest of Brazil, the archipelago of Indonesia, the mountains of East Africa, the South African cape, Mexico, Central

America, the Caucasus Mountains, the Philippines, the west coast of South America, Southeast Asia, the Mediterranean, and on.

Central America consists of only half of one percent of the world's land surface and has cut down over 90 percent of its original old-growth forests, now partially grown back with greatly-fragmented secondary and tertiary forests; subsequently, most of its large array of nearly 3,000 endemic plant species, over 400 species of mammals, and dozens of its endemic amphibians are critically endangered and at great risk of vanishing forever. Many already have.

The Monteverde Cloud Forest Preserve in Costa Rica is one example. As a relatively small part of a mountain range, climate temperatures changed enough in less than twenty years in this very delicate area to cause the protective cloud mists to vanish for the breeding season of the formerly omnipresent, locally-celebrated Golden Toad, which suddenly disappeared forever in 1989, now designated extinct by the International Union for the Conservation of Nature (IUCN).

The story is the same throughout the world: our planet's old-growth forests have been reduced by 75 percent. Besides Russia, Canada, and Brazil, every other country has liquidated 80-99 percent of their forests, including the U.S. at 97 percent and Nigeria at 99 percent. The U.S. has done it for development and profit; the poor and more overpopulated countries have done it for survival, or MNE's have come in as corporate neo-colonialists to extract the forests for their own gain.

Eighty percent of the world's population - 5.5 billion people – lives in the poor and overpopulated countries where many of the hotspots are. Economic desperation motivates hundreds of millions of people to annihilate their forests; for instance, half a billion people carve out subsistence agriculture in tropical forests. Understandably, many don't care what they have to do to the environment in order to eat. Situations are worsened by political instability and corruption, allowing for an anxious rush of illegal natural resource extraction, including animals. Many politicians also perceive their impoverished countries as substandard compared to the wealth of much more economically prosperous countries, and they are willing to sacrifice the unrecognized and underappreciated value of their natural legacies in hopes of benefitting from joining the ways of the admired, dominant, contemporary world.

Additionally, for many legislators and administrators, there is more personal incentive to receive the monetary benefits of exploitative actions rather than to protect biodiversity; of the under 12 percent of hotspots that are protected in any way, much of that is only on paper.

This gorgeous planet, this oasis in the cosmos, was once a lush, unspoiled paradise before it became ubiquitously populated with humans, which now puts enormous pressures on these hotspots. These places are full of rare species, many already critically endangered, which makes them more valuable to people that would exploit them – and exploit them at an ever-increasing rate to get them before they're gone – the Ghost Demand Syndrome (GDS). These areas are susceptible to the ravishment of resources for specialty lumber, minerals, bushmeat, and exotic pets and plants, all extracted illegally on a daily basis. Hotspots are threatened by human encroachment, assaulted by increased human activities, and invaded by the non-native species that follow.

Our world's priorities must change drastically and immediately if we are to preserve Earth's biodiversity. Poor countries don't have the means to protect them; it must come from the countries that have the wherewithal. We must realize that spending billions to protect life is more important than spending the combined trillions in annual international military budgets. What are we protecting ourselves from if doing so increases the desertification of our Earth half devoid of its DNA, eventually replaced with weed species and genetically-distorted freaks of nature caused by anthropogenic chemicals and radiation that was good for the war-based economies of a self-absorbed species that never considered its effect on the nonhuman life of our planet, which meant nothing to them and had no value to them unless it made profits.

EARTH SYSTEMS

OUR EARTH HAS AN iron core, most of which is molten, and as our planet rotates, this combination causes the Earth to become a dynamo, generating a large, healthy magnetosphere that surrounds and protects us. This force field is shaped like a torus, a three-dimensional figure eight, much as how iron filings line up two-dimensionally around a stationary magnet, and it is further sculptured by the solar wind. Our Earth is a spinning electromagnetic turbine and forms an energy field extending into space an average of 50,000 miles from the Earth's surface, shielding us by deflecting the solar wind radiation emanating from the Sun. Without this, our oceans would be vaporized and blown away; obviously, our magnetosphere was and is primary to forming and maintaining life on our planet.

All of our solar system's four inner planets have iron cores, but the creation of oceans on Mercury, Venus, and Mars did not occur partially because they did not have strong enough magnetospheres. For a magnetosphere to be generated, a planet must have an iron core that is at least partially molten and have a planetary rotation fast enough to cause a turbine effect. Mercury has no atmosphere because the Sun's solar winds are more forceful the closer to the Sun a planet is; consequently, Mercury's already-weak magnetosphere (92 percent weaker than Earth's) was overpowered by the Sun long ago, so Mercury has been forever stripped of any gases, including water vapor. Venus, though with a molten core, rotates too slowly to generate the turbine effect needed to create a magnetosphere. Venus is also 60 percent larger than Mercury and 46 percent further from the Sun, allowing its gravity to hold an atmosphere, but it is mostly CO_2 due to the emissions of constant and massive volcanic activity. Mars does not have a magnetosphere because its once-molten iron core has become completely solidified, as it is too far from the Sun and too small (53 percent the size of Earth) to retain the needed heat. Our moon has no magnetosphere: its core is now solid, and, of course, it does not rotate. The four outer giants of our solar system – Jupiter, Saturn, Uranus, and Neptune - all have magnetospheres, but they are mostly gaseous planets unable to contain oceans.

The prerequisites for life to exist on Earth are a partially molten iron core and a fast enough rotation to generate a strong magnetosphere in order to contain an atmosphere and the oceans. Additionally, our planet had to be in the range of the circumstellar habitable zone (CHZ), known as the Goldilocks zone, because Earth's distance from the Sun is just right for the liquefaction of water, allowing us to have a climate compatible to life.

The climate of the Earth is affected by many complexities. Milankovitch cycles are peculiarities in the Earth's solar orbit that take 100,000 years to complete. Other cycles involve the tilt of the Earth's axis relative to the Sun varying from 21.8° to 24.4° over a 41,000-year time span combined with the Earth slowly wobbling in two cycles of 19,000 and 23,000 years from the gravitational influences of the Sun and Moon. Earth's tilt is now at 23.5^0 and is why winter and summer are opposite of one another in each latitudinal hemisphere. Furthermore, the period of precession, which lasts about 25,700 years and governs which hemisphere is at aphelion (further) or perihelion (nearer) to the Sun determines the mildness or intensity of the seasons in each hemisphere. Currently, Earth's aphelion peak occurs thirteen days after summer solstice in the Northern Hemisphere when the Earth receives less total solar energy, and the Earth is at perihelion just after the Southern Hemisphere's summer solstice when the Earth receives more total solar energy. This presently makes winters warmer in the North than in the South and summers cooler in the North than in the South. This circumstance is gradually reversed about every 13,000 years, which will take about another 4,000 years to reach its alternate apex. Add to all this the possible influence on the Earth's climate of the Sun's 11-year cycle of sunspot activity and its 22-year cycle of magnetic reversals along with the elliptical orbit and 29-day cycle of the Moon's governing of the Earth's tides, and that's only partially what we know of the complications that determine the Earth's climate and other functions.

On a more terrestrial orientation, one primary force upon the Earth's climate is the Coriolus effect, which is caused by the Earth's continual rotation and determines whether the oceanic and atmospheric currents of each latitudinal hemisphere flow clockwise or counterclockwise.

The other most powerful influence is the thermohaline cycle which acts as a giant conveyor belt circulating cold, deep, and highly-saline waters throughout the oceans of the world with warm, less salty surface waters. Also known as the thermocline, it has a profound effect upon the ocean's chemistry by taking carbon, phosphate, and nitrogen from the surface waters. Therefore, the thermocline deeply affects the ocean's biology if disrupted; inversely, the ocean's biology can deeply affect the ocean's chemistry. The thermohaline circulation takes a thousand years to fully complete, and its chemistry affects our climate with the ability or inability of the ocean and its lifeforms to absorb and fixate carbon and other elements.

Our Earth measures twenty-six miles more around the equator than it does by measuring around the poles, and it has a 720-mile-thick, solid-iron core with a temperature of $3,000^0$ C. Surrounding this core is 1,380 miles of molten, iron-laden Earth topped by the mantle, which is 1,745 miles thick. The outer 110 miles of the mantle consists of the lithosphere, the surface crust, and the more elastic asthenosphere which the continents drift upon.

As part of the lithosphere, supercontinents break apart and come back together about every 500 million years according to what's called the Wilson cycle. The Earth creates new rock as it spreads from its mid-ocean rifts from seven to eight feet every seventy-five years, the same pace as a human fingernail grows. This causes tectonic subduction at the margins of the continents and is why mountain ranges are pushed up near the coasts. The oldest areas of a continent are in its interior where tectonic subduction has not taken place in some parts for two to four billion years; in contrast, the oldest rocks on the ocean floor have been dated at just over 200 million years old.

Our knowledge of continental movement goes back to over 700

million years with the supercontinent named Gondwana. Gondwana consisted of what are now Africa, South America, Antarctica, Australia, India, and Arabia. After moving southward, Gondwana did a long-term dance swirling over the South Pole for millions of years. Parts of it then stretched northward and meshed with the rest of the continents around the equator and became an even larger supercontinent named Pangaea about 450 million BP right when life was just starting to migrate onto land from the ocean. Pangaea began to break up about 200 million years ago, continent by continent, until the recognizable form of our modern world was complete with India's attachment to Asia at 50 million BP. Presently, this culminates in 68 percent of land mass being north of the equator and 32 percent in the south; this is inversely true for the ocean. It is thought that circa 250 million years from now the Earth's land masses will be formed into one supercontinent again. Proof of the theory of plate tectonics did not come into full acceptance until about 1970, and it revolutionized the earth sciences; with so much interconnected research to it, new evidence is even yet changing details or solving discrepancies.

A very powerful system that affects the climate of Earth involves the Intertropical Convergence Zone (ITCZ) and the Hadley Cell. Since the Sun is more or less directly over the equator, it evaporates huge amounts of water which rise with the heated air, condense, and fall as massive downpours of rain in the equatorial regions around the world. Air temperature decreases in altitude by about 19°F per mile, and it heats back up as it descends (known as the environmental lapse rate). Once the air has lost its water vapor (rain) and been cooled, it is consistently pushed by the rising air behind it; this dry, cooler air is then pressed back down around the latitudes of 30° north and south, becoming hot again, parching the land, and creating the world's deserts on both sides of the equator. The surface air then flows back to the equator to repeat the cycle. Called the Hadley Cell, this gigantic convection loop circles the world; it is also heavily influenced by the Coriolus effect. Hurricanes, known as typhoons in the Pacific, develop when heat is transferred to the atmosphere by ocean evaporation; "the warmer the ocean's surface, and the further down that warmth is, the more powerful the hurricane."[1] Typhoons have been increasing dramatically in size, strength, and occurrence in recent years as a result of oceanic warming.

Another important Earth system is the exchange of carbon between the atmosphere and life on land and the exchange between the atmosphere and the ocean. Carbon has cycled in and out of the land masses, the oceans, the atmosphere, and other lifeforms throughout the eons before it was ever assembled to form our bodies and the book you're reading. Carbon is the fourth most abundant element in the Universe, and 99.99 percent of life on Earth is based on it, all moving within a carbon cycle. The atmospheric/terrestrial cycle, which exchanges close to 60 billion tons of carbon a year, is balanced by the transpiration and decay of plants and the photosynthetic absorption of CO_2 by forests. The atmospheric/oceanic carbon cycle involves about 90 billion tons per year. These two cycles are different because the forests contain over 600 billion tons of carbon, while life in the ocean contains only about 3 billion tons; therefore, the net amount of carbon absorbed by the ocean is much greater than the amount it gives to the atmosphere, and the ocean also absorbs 50 percent more carbon than biota on land to begin with. "[O]n average, a carbon atom will stay in the atmosphere for five years and in the oceans for 400 years. The oceans currently store around 60 times more carbon than is in the atmosphere and 20 times more than in the ecosystems and soils on land."[2]

Many things happen to CO_2 once it is in the ocean. Some of it dissolves and forms carbonic acid (H_2CO_3) or carbonate ions (CO_3), and some is retained within different cycles for thousands of years or more as calcium bicarbonate in shelled creatures, or becomes ocean sediment, or is caught in the thousand-year thermohaline cycle, or as sediment that gets dissolved, or by photosynthetic conversion to organic lifeforms, or by tectonic recycling with continental subduction (which comes back to the land with risen land structures), or to the air with volcanic eruptions - eruptions of less than 2 percent (about 50) of Earth's 3,000 potential volcanoes per year.

Lastly, the Earth's water system involves fresh water evaporating from the saline ocean and falling upon land as freshwater rain. About 50 percent of the Earth is covered with clouds at all times. Rain continually leaches mineral salts from the rocks, flowing salt back into the ocean.

Salts become sediment on the ocean floor and later become new rock from tectonic subduction over hundreds of millions of years. "This recycling... keeps the saltiness of the sea constant."[3] Fungus, algae, and lichen also play major roles with water in the erosion of rock and the salinization of the ocean.

With all these various cyclic interactions, the Earth is far from ever being static. In fact, with the addition of the evolution of life and its impact, there never has been and never will be a time when the Earth experiences the same circumstances it once did even if all the inanimate cycles come together in any previously similar combination. There will never be a time when Earth returns to a previous condition. Humans have already made an eternal impact upon the Earth, and we have done it with unconscious, reckless, gluttonous narcissism overshadowing all of our amazing accomplishments - or vice-versa.

ATMOSPHERE

THERE ARE FIVE LAYERS to our atmosphere, all with diffused boundaries. Three-quarters of atmospheric gases are in the lowest layer, the troposphere, from the Earth's surface to about four miles high over the Polar Regions and about ten miles in altitude over the equator. This stratum has the highest density of gases and is named for its weather-producing turbulence.

Above that is the stratosphere, which rises to thirty-one miles in altitude. It is this layer that contains the Earth's ozone protection, without which life could not exist as we know it, for Earth would be radiated by lethal doses of ultraviolet light that would kill everything it reached, including bacteria. Although the hole in the ozone mostly occurs over Antarctica during its spring month of October, it is still being depleted by the use of halocarbons, which include chlorine, bromine, and fluorine. At the air pressure of Earth's surface, our ozone layer would be less than only four millimeters thick.

The mesosphere, from thirty-one to fifty miles in altitude, is where incoming meteors start to burn up.

Next up is the thermosphere, fifty to 400 miles in altitude, the layer that contains the ionosphere. Ions from the solar wind interact with the magnetosphere here and produce the gorgeous lights of the aurora borealis in the far north and the aurora australis in the far south. It is also in the thermosphere where Earth's gravity begins to release its hold on matter.

Above all these layers is the exosphere where any atmospheric gases finally disperse enough to disappear into outer space up to 40,000 miles from Earth's surface.

If the atmosphere was compressed to the density of water, it would only be thirty-three feet deep (compared to the average 12,514-foot depth of the ocean).

Only two-billionths of the sun's total energy reaches the earth, but its infrared heat is enough to create all the convectional motion in the atmosphere and the ocean once that infrared energy hits matter. Additionally, our planet is 63°F warmer than it would be if not for the Earth's naturally occurring greenhouse gases; without these gases, the temperature of the Earth's surface would be about the same as the Moon and Mars (⁻4°F) rather than its average of 59°F.

Our Earth is the only planet in our solar system with nitrogen and oxygen for an atmosphere. It is life on Earth that also makes a tremendous difference, for not only has cyanobacteria created our oxygen, life has concentrated carbon into many forms for up to 2 billion years, eventually becoming sequestered away as gigantic deposits of coal, natural gas, oil, peat, and limestone.

FORESTS

A FTER FUNGI HELPED ALGAL cells take hold on land and develop into plants about 420 million BP, the first trees, lycopods, came into being around 30 million years later. The overabundant success over tens of millions of years caused the self-inflicted demise of lycopod trees, and gymnosperms (conifers) took their place near 18.00 GY, 280 million years BP, just 1.17 galactic orbits ago. Conifers have been greatly successful ever since, starting in the time of Pangaea.

There are three basic types of forests in the big picture: tropical, from 0^0-30^0 north and south latitude that grow the fastest; temperate, from 30^0-60^0 latitude, which grow the biggest; and boreal, which grow the slowest and sparsest, from 60^0 until it's too cold both in latitude and in the altitude of mountain ranges around the world. There are tropical rainforests and temperate rainforests, tropical dry forests and temperate dry forests.

More than half of all terrestrial species are dependent upon trees. "The biomass of life on land is hundreds if not thousands of times greater than the biomass of life in the seas. Much of this massive presence, an estimated 84 percent, is taken up by trees."[1] Trees were and are the homes of our primate ancestors; they were our safety and are still highly revered by many. Trees also regulate the water cycle of their local environments.

Our reverence and awe of trees was replaced by our construction of pyramids and temples that rose even higher to the gods and aligned with the stars, later replaced by churches with tree-like spires exalted to the God of one's choice. These were replaced by skyscrapers even higher when economics became the secular god, and trees became a commodity of "green gold," creating fortunes for some as valuable as "black gold" (oil).

Since the spiritual veneration of trees, temples, and churches was replaced by economic fanaticism, only 25 percent of Earth's original, pre-human forests are left, unbroken only in the remaining rainforest of the Amazon and the boreal forests of Russia and Canada. Otherwise, in all countries around the world, up to 99 percent of all original forests have been destroyed, including in the U.S. One thing forests all have in common is the DC has always had an obsession with cutting them down. The DC's fervor over economics has supplanted the ancient forests, finding no spiritual value in them at all. The enchantment of ancient forests, our prehistoric roots that cultivated a distinctive, intrinsic reverence within us, is all but gone, lost to the world of industrial commerce and anthropogenic artificiality.

Historically, tropical forests covered 14 percent of Earth's land surface, which equaled the size of all of the U.S. and China combined. Now down to 5 percent, our world's tropical forests have been reduced by two-thirds to a size equal to about 75 percent of the lower forty-eight states of the U.S. Without change, in fifteen years, 80 percent of what's remaining will also be gone, making 93 percent of the world's tropical rainforests destroyed. The problem is, tropical rainforests don't grow back; they're inundated by people, roads, cattle, farms, cities, invasive species, and catastrophic soil erosion. The loss of species is also cataclysmic since tropical rainforests support the majority of all of Earth's land species. For example, Great Britain has "1430 species of flowering plants and 35 native tree species, whereas the Malay Peninsula, with only half the area of Great Britain, has 7900 flowering plants [five and a half times more] and 2500 native tree species [over seventy times more]; of the 9040 bird species in the world, almost half live in the Amazon Basin or in Indonesia; and studies in Peru found 300 tree species in a 2.5 acre plot...."[2] Our world loses a minimum of 40 million acres of tropical forest per year, ten times the size of all the Hawaiian Islands. In twenty years, that's a cumulative area larger than the size of India that is cut, burned, and permanently cleared away for other uses. At least twenty-five percent of our world's species will go extinct just from deforestation.

Unique to tropical forests are the issues of rapid human population growth and the predominance of "[p]overty, corruption, abuse of power, political instability, and a frenzied scramble for quick riches [that] are common denominators of the... developing countries around the world."[3] This leads to such survival tactics as slash-and-burn subsistence farming, illegal logging and harvesting of other plants and animals, and bush-meat consumption. The boundaries of whatever parks that have been set aside are ignored, and protection of species and resources is non-existent. Briberies go a long way in places that feign protection on paper only; the creation of park boundaries has allowed officials to become underhandedly rich. "On paper, Brazil maintains 30 nature preserves in the Amazon, but... together employ a total of 23 guards. [In comparison,] the United States National Park Service employs about 4,000."[4] The creation of parks and preserves peaked in the late 1980s; little has been done since, as economic interests have taken over, and some countries' leaders are embarrassed by still being regarded as "undeveloped," making their wildlands seen as a hindrance and of little value.

John Terborgh, who spent his entire career in the tropical forests says, "What I have seen convinces me that the conventional wisdom now being applied to the conservation of tropical nature is misguided and doomed to failure. ...Nearly every park in the tropics is being degraded...there is nothing left of the park[s] but [their] name."[5] The same "conventional wisdom" is still being applied. For instance, "Indonesian forestry law is violated with impunity: home to critically endangered Sumatran tigers, elephants, and orangutans.... Approximately 65% – 80% of Indonesia's timber production is illegal...."[6]

In the Amazon, the European demand for mahogany starts the cycle of destruction of the rainforest. Mahogany is illegal to cut but is worth $30 to the fellers; it increases up to $3,000 at the nearest sawmill where it is then sold to an exporter who hauls it to a port and sells it for up to $50,000. Depending on what it gets made into –furniture, guitars, coffins, etc. - its value rises to over $250,000 for that one tree.[7] The damage done to access even one mahogany tree is irreversible. In the overpopulated country of Brazil, it first opens up a route to desperate subsistence farmers. Their impact then leads to larger and larger uses of land, culminating in mechanized agriculture such as miles of feed lots, sugar cane or soya fields. Ultimately, this leads to severe desertification in the former Amazon rainforest. In a time of only twenty years, the authors of *The Last Forest: The Amazon in the Age of Globalization* witnessed this progression, stating that "…[t]he stampede into the jungle… had… cut and burned [all the trees] to make way for pasture… grass sown in its place had grown sparser each year until it quit growing altogether. The desert of failed ranches went on for an hour – gray sky, gray dirt. It was 50 miles of moonscape."[8]

As well as Brazil, the Amazon rainforest includes parts of Columbia, Peru, Bolivia, Ecuador, and others, covering an area "larger than the size of the continental United States west of the Mississippi." The Amazon River is 4,000 miles long and up to 200 feet deep in places. It rises and falls 65 feet from the rainy to dry seasons, its daily flow is eleven times that of the Mississippi, and its delta swells to *two hundred miles wide* at the Atlantic. It is the richest area on Earth in biodiversity; perhaps 20 percent of Earth's species live there, many endemic.[9] Before the rainforest was logged in parts of western Brazil, sixteen hundred types of butterflies were recorded, but species were reduced to fifty from one forest fragment to another afterwards, a 97 percent drop.[10] "The Amazon rainforest accounts for 10 percent of the carbon stored among the world's ecosystems…"[11] and "contains three to five times more carbon per acre than an open, dry forest; an acre of [the Amazon forest] up in flames equals three to five acres of Yellowstone."[12]

The loss of tropical rainforests goes on. By 1996, Ecuador had lost over half its forests due to human economic activity, including the loss of over 3,000 plant species to extinction since 1950[13]; by 2009, the 820,000-acre rainforest of the Monte Azules Bioreserve in Mexico had lost one-quarter of its forests to illegal logging.[14] Mexico stands out for its rich biodiversity and endemism, having more reptiles than any other country on Earth.[15] By 2009, Central America had lost 90 percent of its old-growth forests, which held 8 percent of the world's biodiversity in its half of one percent of Earth's land surface.[16] By 2006, the Philippines had lost over two-thirds of its lush rain forest, lost to logging by MNEs, agriculture, and open-pit mining.[17] It goes on and on.

Everywhere the DC has been the trees have vanished, mostly for agriculture and development to meet the demands of the expanding human population. "During the past two millennia deforestation became severe first in the temperate countries. It spread from the Middle East and Mediterranean to Europe, thence to Northern Asia, and on to North America. Finally, in the twentieth century, forest destruction swept through the tropics."[18] "[Centuries] ago, 90 percent of the Highlands [of Scotland] was covered with forests, but logging, sheep grazing, and grazing by red deer [who ate all the tree seedlings after all the wolves were killed, which allowed the population of deer to explode] have almost eliminated the trees entirely."[19]

In America, getting rid of all the original trees has always been the dominant forest policy; national forests are even managed under the USDA, the Dept. of Agriculture, because they're seen as a crop, a commodity. Since the twentieth century the DC has only regarded old growth forests as non-productive crops that are wasting precious land space; it sees only money and cares not about the species that need the forests to live. The DC's mental disease has put forests into a controlled and orderly nature and turned them into single-species tree farms, having no problem at all with erasing the old growth. There is no original forest remaining east of the Mississippi; it's long gone but for a very few, extremely small patches. "The most diverse temperate forest in the world blanketed the land from the Atlantic to the Great Plains. White pines and tulip poplars grew to over 200 feet in height and individual chestnut trees spread their branches over a quarter of an acre."[20]

In just one generation, starting in the 1870s, the entire virgin white pine forests of the North Woods of Wisconsin, Michigan, and Minnesota were leveled. It was here that the fastest, most voracious logging destruction ever known at the time gave rise to the bigger-than-life Paul Bunyan, a mythical folk hero still mindlessly celebrated today for the DC's egotistical triumph of waste, progress, and perceived power over nature.

As the DC moved west, the American government gave the railroads twenty square miles of land (12,800 acres) for every mile of track laid from Minnesota to the Pacific Coast, totaling about 130 million acres. The Northern Pacific Railroad received 39 million acres of free land, did some further trades of "worthless" land for more valuable land, and eventually sold nearly a million acres of prime, Pacific old-growth rainforest to Frederick Weyerhauser for $6 an acre in 1899. The Weyerhauser logging company later became the largest private timber owner in the world with a total of 13 million acres in the U.S. and Canada. Other logging behemoths, such as Potlach and Boise Cascade, started the same way and proceeded to decimate the ancient forests of the Pacific Northwest, untouched since the glaciers receded from the Ice Age. These giant forests were hacked down equal to the rapaciousness of tent caterpillars or locusts, forests with three hundred-foot trees of Sitka spruce, noble fir, Douglas fir, yellow cedar, red cedar, hemlock, Pacific silver fir, and more, almost all as big and majestic as the redwoods and perhaps the most gorgeous, awe-inspiring forests ever to exist on Earth, spreading from northern California to Glacier Bay in Alaska, from the beaches of the Pacific Coast to the coastal mountain ranges, full of wildlife and the purest of waters abounding with salmon, even up to just forty years ago.

While the attention was on the Amazon forests in the 1980s, the temperate forests of the U.S. were being liquidated; the Reagan administration had over 40,000 miles of logging roads built during its eight-year reign, heavily subsidized the timber industry, and sold the massive old-growth trees to the big logging companies for a price equal to a six-pack of beer. Logging went into high gear. Ninety-seven percent of Oregon's old growth is now gone, and the numbers are similar in Washington, British Columbia, and California – and the industry is *still* after the rest, angry at not being able to have what little has been preserved in parks and wilderness areas. During this time, the logging industry convinced many that the lack of jobs (due to the eventual lack of trees to cut) was the fault of environmental preservationists, ignoring the fact that over one billion board feet of raw logs per year, just from the Olympic Peninsula alone (of Washington State), were being exported to Asia, more than fifteen times the number of board feet that were staying at home, and that new advances in technology were replacing workers, even while the big lumber companies cut workers' pay in spite of making net profits of $1 million per day.

The trunks of the thousands-of-years-old Pacific Coast ancient forests used to have to be cut lengthwise in quarters to fit onto the logging trucks to haul to the sawmills or ports. Now the trees only grow to about sixty years old before harvest – 2 percent of their previous age – and twenty trees can fit on one truck. The ability to cut the forests down has greatly accelerated with advanced technology. Feller-buncher machines can literally cut through a tree in a second and then hold and carry an upright bundle of trees to a pile. Such expensive machines need to be run full time to keep up payments to be cost effective, so the needs of economics supersede the needs of sustainable forestry.

Incredibly, this being the case of the temperate rainforest, "Arizona has lost more of its virgin forest [temperate dry] than has any other western state."[21] About 98 percent of Arizona's magnificent old-growth Ponderosa pine, pinion pine, and native spruce are gone; almost two hundred years of intense overgrazing and the extermination of predators and prairie dogs has befallen; and the resultant ubiquitous takeover of invasive weed species have all combined to the point of destroying the Arizonan ecosystems with such drastic oversimplification that any of its former biological diversity is virtually all deceased, with ghost species all that remain. What was an arid state to begin with has become fully desertified and full of weed species where its former forests once existed. This is true throughout the Southwest.

There are no words to describe how disastrous clearcuts are. It's no different than the aftermath of a nuclear bomb explosion without the radiation, but the damage is in squares and rectangles. Every tree within thousands of acres has been cut down and entire mountainsides and valleys stripped naked and drug off (in British Columbia, Canada, clearcuts have been as large as *square miles*). What remains of gorgeous and diverse green forests, sparkling in the sun and full of life a day or week before, are the gray-brown splinters of "weed" trees strewn chaotically about like giant, broken toothpicks among the gray-brown stumps, forlorn branches of skinny, mangled shrubs dangling a leaf or two from limbs protruding up from their graves. Streams and rivers are walked away from choked with debris and brown silt. Tank-like tracks of giant bulldozers remain with spills of hydraulic fluid soaking into the freshly churned soil, crumpled oil barrels and abandoned choker cables left to rust. The land of a clearcut is left where even the smallest of animals no longer exist; land where the wasted slash has been burned with synchronized explosions of napalm in attempts to send the smoke up to the stratosphere, the fires often getting out of control and burning healthy forest; land where even the humus, crucial mycelium, and soil bacteria are burned away in man's greedy ignorance and economic expediency; land with nothing living to absorb the rains that turn to avalanches of mud that kill the fish and cause floods and road washouts; land suddenly without a hint of cool shade getting baked in the sun, soil drying out in a rainforest zone, heating the rivers and streams, killing riparian species. Whole checkerboard swaths of once Eden-like land are left in a state of absolute ruin, eventually replanted usually with only one species - genetically-modified Douglas fir – and only because the law finally forced the industry to adopt a policy of minimal restoration some forty years ago. This is an example of a pattern now found around the world.

The stark contrast between the pristine life of a forest and utter death is no more apparent than the miles of border separating clearcuts in the national forests surrounding the intact forests of National Parks. In an extreme example of islandization, timber companies logged completely around Mt. Rainier in Washington State in a successful effort to prevent any possible future expansion of the park, leaving a barren wasteland right up to the straight boundary lines of the park, perhaps even as an affront to anyone outside the industry. This is referred to as "panic clearing" – another example of the Ghost Demand Syndrome – when industry logs rainforests as much and as fast as possible "in the fear that the government [will] prevent them from doing so in the future."[22] "The industrial landowners... are in business [only] for profit. ...Social and ecological considerations are... [not their] primary economic goals."[23]

There are many examples of the industry's mindset, the mindset of the DC. Charles Hurwitz, as if he believed in God and the bible, stated that Genesis gave him the directive from God to dominate and subdue the earth. After his Maxxam Lumber Co. purchased the vast redwood forest of the Headwaters of Northern California from the Pacific Lumber Company with junk bonds (higher corporate interest rates for loans to risky companies), he directed a frenzied policy of deforestation in the early 1990s to pay off the junk bond interest rates.[24] The loss of live, ancient forests full of biodiversity and centinelan species for manmade *interest rates….* This policy escalated when the government decided part of the Headwaters were finally to become protected habitat. Maxxam then began logging giant redwoods twenty-four hours a day every day and night for months until the legislated protection took effect, using artificial lighting and generators for night operations. Another timber DIC CEO said, "I don't think the environmentalists realize the reason the Lord created the resources. He created the trees for us to use. Timber is a crop…. Not to use it would be a waste of a material that we would never be able to reclaim." A logger in an ad said, "What people don't know is that I'm clearcutting to save the forest – the same way nature does."[25] Cafes in the small towns of logging areas had signs that read, "Spotted owl burgers served here," or in Fourth of July parades chased mock spotted owls with chainsaws. Others went out and hunted spotted owls to kill before scientists could find them and protect the habitat for this flagship, endangered species *and the thousands of other lifeforms of the ecosystem the owls represented.* Companies state they are cutting the forests to prevent nature from letting them go to waste and then use double-speak to call it conservation. One representative of the infamous Bowron Clearcut in British Columbia, which ultimately covered an area of 625 square miles (25 miles by 25 miles), tells shocked observers, "What you are seeing, folks, is not the world's largest clear-cut, it is the world's largest tree plantation!"[26] The industry has openly declared its intent to clear nature's forests and transform them into tree plantations. Yet other corporate spokespeople state that "…if what you see looks degraded and ugly, the fault lies with your vision rather than [our] company's management. Once you learn the proper attitude, it should look all acceptable, if not admirable."[27] Even the famed Thomas Jefferson said the enormous American land mass contained "great quantities of land to waste as we please." Using the typical denial strategy of endless delay while lucrative operations progress, since court costs and environmental fines are miniscule compared to profits, the timber philosophy states, "If an activity cannot be proven harmful, the activity shall be carried out."[28] "The disgraceful condition of industrial forestland is sufficient to disqualify corporations from owning land" at all.[29]

Redwood forests once dominated our planet, as well as other

giant conifers, including on what is now Antarctica, which was part of Pangaea at the time. The asteroid impact 65 million years ago had no major effect upon redwoods, which the dinosaurs lived among, and redwoods were still flourishing during the dawning of the Age of Mammals 55 million years ago. As well as Antarctica, fossils of redwood forests have been found all over the Northern Hemisphere from islands in the Arctic Ocean to near the equator, including Siberia, France, Columbia, Wyoming, and on.[30] At 140 million BP, redwoods began sharing the planet with flowering trees, the angiosperms.

The virgin temperate rainforests where trees grew over ten times larger than anywhere else in the world have all but vanished from our Earth along with the species that had nowhere else to go. The only places where intact temperate rainforests of old-growth still exist outside of the fragments of parks are in Russia's gorgeous Primorye, home of many endangered species in the Far East, including charismatic megafauna; Southeast Alaska; and smaller areas in Chile and Argentina. They amount to a combined total of less than 3 percent of the area covered by the remaining tropical rainforests; "the temperate rain forest then is one of the rarest, most severely threatened forest ecosystem types in the world."[31] Home to critically endangered species, it is pure senselessness and greed not to preserve and protect these forests immediately by any legal means necessary, such as World Heritage Sites, National Park status, wilderness areas, or other. "The World Bank estimates that half of all logging in the Russian Far East is illegal,"[32] and only *thirty-five* wild Amur leopards in the world exist there.

Canada is the second-largest country in the world and contains almost ten percent of the world's forests. "Canada's boreal forest is almost equal in size to the Amazon rain forest," and with its massive, legendary clearcuts, "it is known as 'the Brazil of the north.'"[33] The fragile boreal forest is made up of nearly 50 percent aspen trees; once burned as a weed species, aspens are now valued for their superior pulp fiber, and nearly every tree is destined to become paper pulp in contracts made with giant MNEs you'd never suspect, such as foreign car manufacturers.

Canada's logging industry also produces fewer jobs per volume of wood cut than any other industrialized country[34], largely because of mechanization such as feller-bunchers, which were practically made for the smaller boreal trees and sparse underbrush.

Nordic countries' have also been liquidating - clearcutting – their ancient forests, causing the extinction of over 200 plant and animal species and adding another 805 to the endangered species list.[35] In Russia's Far East and Siberia, the environment is in a tailspin dive, as people are doing anything they can to survive in a lawless, economic free-for-all, and the government enforces no environmental laws there.

Deforestation has a direct impact on increasing global warming. As enormous amounts of carbon are released from burning forests and the slash leftover from logging, and a lower albedo (for definition refer to index) raises surface temperatures without the shade of forests and dries out the soil, a positive feedback loop causes a pandemic outbreak of bark beetles that kill standing, drought-stricken forests, which then cause massive forest fires. Alaska's average temperature has risen over 4^0 F in the past fifty years, and the entire forests of the Kenai Peninsula have burned away. There are over 500 living species of conifers; one in four is threatened with extinction.

"In the space of a century, we have wrought a holocaust in the wilderness. And we continue." – Dave Foreman[36]

Turning wild forests into artificial (GMO), monoculture tree farms and spraying them with pesticides and herbicides has and is completely undoing the complexity and stability of life in the forests and our planet. Timber managers haughtily thought they could improve an ancient forest system before they even tried to understand it; in-depth studies didn't even begin until the 1980s. The Forest Service was ever only organized primarily to log and sell timber, using the term "management."

Results from some of the studies of the past thirty years in the old-growth forests have to do with mycorrhizae and lichens. There are over 5,000 named types of mycorrhizal fungi that are plant symbionts[37]; they are intertwined among the root hairs of species from 90 percent of all vascular plant families. Over 80 percent of all plants would die without mycorrhizal fungi, especially trees.

Lichens are an interdependent blend of bacteria, fungi, and algae; they are classed as part of the Kingdom of Fungi, are very slow-growing, and have the important role of fertilizing forests by fixating atmospheric nitrogen to soil. When a clearcut takes place, all the lichen of the area is destroyed; for instance, the Loberia lichen, particular to temperate old-growth forests, takes two centuries for it to even be seen again.[38] A forest of Douglas firs and others in the Pacific Northwest rainforest may be older than the U.S. but still be uninhabited by Loberia. It takes 800 years for biodiversity in an old-growth temperate forest to reach full maturity. Ninety-six percent of the old growth forests in the Pacific Northwest have been logged away; even in an ideal situation, they would not return to true old-growth status until about 2800 AD[39], and that would be without the unknown number of centinelan species already gone to extinction.

ISLANDIZATION

I SLANDIZATION IS A TERM used to describe severely fragmented areas of terrestrial ecosystems that have become like islands in the ocean due to the encroachment of human activities. This occurs primarily due to deforestation, farming, warfare, state and national boundaries, the establishment of parks and preserves, urbanization, and road building. This also gives rise to easier access for increased recreation and illegal hunting, all of which causes isolated species to become particularly vulnerable to extinction.

In actuality, oceanic islands have the highest rates of extinction on Earth, plus they usually contain the greatest number of endemic species. Hawaii has the highest amount of extinct island species in the world with only two remaining native mammals out of its original forty, and both are critically endangered; over 200 hundred (74 percent) of Hawaii's largest native plant species, the lobolioids, are already extinct or on their way out; over half the native birds out of 140 that were known are now extinct; and that's only mammals, birds, and one grouping of plants.

Madagascar, Earth's fourth-largest island at twice the size of Arizona, contains all of the world's lemurs, thousands of endemic plant species, and about 200 types of frogs; most are critically endangered, since over 95 percent of the island's forests have been clearcut. From space-satellite photography, Madagascar, with its reddish soil, once long appeared to be copiously bleeding far into the Indian Ocean with the logging's resultant soil erosion.

Then there's Guam with its non-native, omnipresent brown snakes that have killed off every bird and bite children in their sleep. Islands around the world all have their related problems; even Australia, the "island continent," leads the world with far more extinctions than any other continent (excluding Antarctica), having the highest rate on Earth of extinguishing mammals and other species.

However, Singapore, the city-sized island country lauded for being 99.75 percent urbanized, lost most of its fish, mammal, plant, butterfly, amphibian, and bird species to extinction, with half the remaining as mere relics.[1]

All this is to say that it is the same with the islands of wild habitat humans have created on continental land, and the threats to these islands of fragmented habitat are numerous. Most parks and nature preserves are not large enough to support apex predators, which changes ecosystems dramatically. The meso-predators (mid-level) then proliferate and kill off more of the birds and smaller animals. Even trees and other plants become what biologists call the living dead – ghost species - as the overabundance of elk, deer, and rats, for example, eat the young seedlings, which never grow to replace the dying off of mature trees. Most islandized lands are so small and vulnerable that a single natural catastrophe, such as fire, can destroy them and the inherent species. As ecosystems migrate with continued climate change, species within stationary islands such as national parks lose their habitat and have nowhere to survive or go. "[W]ith so many species on the move, a reserve that's fixed in place is no stay against loss."[2]

Parks and preserves make for much easier targets where poachers know it's where they'll find the endangered, high-priced game animals or trees such as mahogany and teak. "In Africa and Asia, bioreserves have become the preferred hunting grounds for poachers and bush-meat traders; it is, after all, where the animals are."[3] As human encroachment closes in, whole towns exist right alongside parks and preserves where wild animals try to exist right next to humans. Most of these situations are in countries where lack of funds, corrupt officials, or political unrest makes wildlife-protection laws nonexistent. For instance, in Central Africa, at least 1 million tons of wild-animal "bushmeat" is consumed per year and is "80 percent of all animal-based protein consumed."[4]

Populations of pollinators also become greatly reduced as "healthy" fragments of prairie, forest, and desert habitat become too few and far between. Most species cannot move from one fragment to another without a corridor of habitat connecting them, which is not a concern of businesses that extract natural resources. The denuded lands around islands of habitat also allow easy access for the yahoo culture of ORVs, ATVs, dune buggies, dirt bikes, and others to intrude upon and further harm compromised land. Islandization is caused by habitat destruction surrounding the splinters of once-gorgeous, contiguous landscapes; it is the number-one cause of isolating species until they finally blink out.

An adequately-sized wilderness area needs to consist of at least 10,000 square kilometers (an area of about 62 miles square or 3900 square miles). This is the minimum required to support a genetically-viable population of apex predators that will maintain the balance of an ecosystem. Extremely few protected areas exist anywhere of this size, especially with the enforcement that's necessary in the overpopulated, more tropical countries where the greatest number of humans and non-human species live. Additionally, a paltry amount of funds are provided worldwide for habitat and species protection compared to what the U.S. provides just for its own National Park Service.

Islandization isolates species. Without enough space for top predators, ecosystems start to go haywire, and up to 75 percent of vertebrate species eventually die off in a trophic cascade. In Venezuela, tropical howler monkeys exemplified this when they became isolated on large hilltops after islands formed within a lake (Lago Guri) that rose behind a new hydropower dam. Their predators had left the rising waters, and the monkeys became overpopulated, denuded the trees, got sick on other plant foods, attacked and maimed each other, and killed their own young. The islands became practically barren, keeping alive the smallest of only 25 percent of the original vertebrates, all the result of human expansion.[5] This is an aggregate syndrome occurring around the world.

PRIMATES

I T IS GENERALLY ACCEPTED that the impact of a giant meteor caused the mass extinction of species 65 million BP, which included the demise of the dinosaurs. After nature is thrown into such chaos as a mass extinction, surviving species revive, evolve, and compete with one another for up to 10 million years until a form of cooperation develops which establishes a new, lasting equilibrium. This 10-million-year recovery aligns with the beginning of the Eocene epoch, 55 million BP, marking the onset of what became the current Age of Mammals.

There were small, nocturnal mammals that had co-existed with the dinosaurs, and after the 10-million-year recovery, some of these survivors emerged as lemurs. Lemurs look like part monkey, cat, and raccoon with large, nocturnal eyes. There are many types of lemurs, but they exist solely on the island of Madagascar, Africa, and they are all classed as endangered species – most as critically endangered - because they are arboreal animals, and the island has been catastrophically deforested by 95 percent to make room for human use and exploitation.

Lemurs are the first primates and led directly to the evolution of humans. In general, though numbers fluctuate with new discoveries, beginning with lemurs at 55 million BP, DNA split from them to gradually form monkeys at about 40 million BP. At 20 million BP, DNA began expressing as gibbons, the first apes, starting the primates' descent from trees. The absence of tails, which greatly benefits tree climbing, is the main difference that separates apes from monkeys and lemurs. From gibbons, DNA then developed into orangutans at 15 million BP and into gorillas at 10 million BP, and the chimpanzee split that led to humans occurred from gorillas at 7 million BP. Chimpanzees and bonobos (formerly called pygmy chimpanzees) then speciated from one another about 800,000 BP. Gorillas are genetically closer to us than they are to chimpanzees or orangutans, and chimpanzees are genetically closer to humans than they are to gorillas; the chimpanzee/bonobo genome is 98.6 percent similar to our own.

Of all the primate species in the world, about 70 percent are monkeys. The other 30 percent is comprised of dozens of lemurs and what are known as lesser apes and greater apes. Of the greater apes, there are only six species: the Sumatran and the Bornean orangutans, the chimpanzee and the bonobo, and the western and eastern gorillas. Altogether, half the world's non-human primate species are facing extinction, and one out of eight are critically endangered. In Asia, 71 percent of primates are said to be at risk of extinction according to the IUCN.[1] "The consensus... is that this is just the beginning of a great wave of primate extinctions...."[2]

The fifty percent of primates threatened with extinction includes bonobos, gorillas, orangutans, gibbons, and lemurs. The wild habitat of gorilla, chimpanzee, and bonobo populations is limited to West and Central Africa - areas which are being decimated by war, poverty, and human overpopulation. All the world's wild gorillas basically live in just seven small countries in Central Africa; Rwanda, for instance, is only as big as the U.S. state of Vermont (or a quarter the size of Ohio). The Western Lowland Gorilla and the Mountain Gorilla subspecies are both critically endangered. Their populations are still shrinking due to the tiresome and standard "business as usual": logging (much of it illegal), mining, oil extraction, poaching and bushmeat consumption, governmental corruption that abandons the protection of wildlife in designated preserves and enables the other abuses, and conversion of forests to agriculture (including livestock) to support the encroachment of more people in their desperation or greed. They are also being wiped out by the horrific Ebola virus and stressed by climate change.

Orangutans are known as the smartest animals on Earth second to humans. But orangutans are rapidly dying off as they lose their tropical, Indonesian forest habitat to palm oil plantations. The orangutan population was historically up to 800,000. Now it is down to only 60,000 and falling fast, due mostly to habitat loss and hunting. The Indonesian government actually paid bounty hunters and others to kill orangutans to make way for clearing the forest to establish the very lucrative palm oil plantations, the trees' fruits which are harvestable every ten days. Ultimately palm oil is used for many things such as alternative fuel, cosmetics, and cooking oil, which also makes the ubiquitous (GMO) corn chips used for appetizers, private parties, football spectators, and cheap potluck contributions, etc., which all contribute to the extinction of orangutans.

While we're obliviously doing all our human activities, and even though they are critically endangered, whole families of innocent gorillas are still being murdered just for a couple body parts, and cute little baby gibbons are being sold as house pets after they are caught before they hit the ground as their mothers are shot and killed in the trees, dropping their babies from their suddenly-limp, extremely long arms in the few disjointed forests that are left.

Chimpanzees and bonobos are both threatened with extinction. The big difference between these two species is that chimpanzees, ten times stronger than humans per size, are known to attack others of their species very aggressively at times, actually to the point of dismembering and disemboweling their adversaries, tearing them to pieces by hand. Bonobos on the other hand, the only apes commanded by females, show very few signs of aggressiveness and are known, rather, for their peacefulness and prolific sexual relations with various mates. Estimates place the population of chimpanzees currently around 230,000 compared to millions only a century ago, and though these numbers do not technically put them in the critically endangered category, it has been recommended to immediately do so since 2008, as their numbers are crashing so rapidly. The reason for their demise is the same as with the other primates, although chimpanzees are more susceptible to being eaten as bushmeat. The much more secretive bonobos have a population of only about 30,000, equal to less than half the human attendance of only one football stadium in the entire world.

NATURE DEFINED

NATURE IS THE TOTALITY of Creation, the transcendent ruling force, animate and inanimate, on and affecting the Earth, fully functional without humans. In the animate world, the world of DNA, all species are driven by the same dynamism: "to reproduce as fast as [they] can and to invade any habitat in which a toehold can be gained."[1] The drive of DNA within humans has caused the end of the wild and the creation of self-serving order: domestication, which has brought progressive artificiality and disassociation from the wilderness - the wild, free, and self-willed.

HISTORY OF CONSERVATION

SOON AFTER THE HORDES came to America from Europe, and the Natives saw how shockingly bad the newcomers behaved towards nature, the Indians started warning them of the dire consequences that would result. The newcomers never listened, brushing the Natives off as unsophisticated, unevolved, uncivilized, and ignorant; their ecological opinions didn't matter whatsoever. This has gone on now for about 500 hundred years.

From 1500, it took a long span of over 300 years before voices from within the DC finally woke up enough to lament about the loss of the natural world with their transcendental writings praising God's amazing creation of nature, first with Emerson in the 1830s, then Thoreau's publications up to the 1860s. Their religious and highly mental exaltations resonated with the puritanical religious like-mindedness within the DC. Their thinking actually continued to separate humans from nature by elevating humankind's unique ability to conceive of and contemplate God, best done in nature, which was essentially the only reason it deserved preservation. At the same time, they proposed it was man's highest calling to transcend his own physical nature by mastering the baseness of incarnation to live a pure and ascetic life. Thoreau, who lived to age forty-four, was a lifelong virgin by choice, an extraordinary separation from nature. The rest of the materialistic DI never bothered to listen, thought the transcendentalists inconsequential and self-indulgent mental elitists, and the juggernaut of a very destructive civilization rolled on at full speed. It was during this time the virgin forests of the Great Lakes states were completely logged off in only one generation's time like the feeding frenzy of piranhas, and the domination of nature was boastfully celebrated by a society that believed it was chosen by God with Manifest Destiny.

During the last half of the nineteenth century, voices of the vanishing Native culture sorrowfully gave out their last wise and prophetic warnings through those such as Chief Se'atl (Seattle) and dozens of others. Still nobody listened; now that the "sub-humans" were defeated, they merely became sentimentally venerated but still dismissed as inconsequential.

More voices arose out of the DC and became activists. John Muir was instrumental in establishing Yosemite National Park, in 1890, and in founding the Sierra Club. Theodore Roosevelt, though a big game hunter, had more lands preserved in the early 1900s and established the Forest Service in 1905 after discovering the forests were being destroyed by 500 percent more than they could grow back. It was not until the 1940s that Aldo Leopold came along with his conservationist writings, helped found The Wilderness Society, and was the first member of the DIC to come up with the "novel ideas" of having a land ethic and to "think like a mountain," as if Natives had not been saying this for 450 years and forever. Still the rapacious DIC didn't listen.

With the arrival of the 1960s, the problems of environmental destruction could no longer be ignored. The World Wildlife Fund was created in 1961; Rachael Carson published her watershed book, *Silent Spring*, in 1962, which was the real launch of the modern environmental movement; the Clean Air Act took effect in 1963; author and activist David Brower helped get the monumental Wilderness Act passed in 1964 and set up Friends of the Earth in 1969; and Congress passed the National Wild and Scenic Rivers Act in 1968. Before the National Environmental Policy Act of 1969, federal agencies did not even have to consider the effects of their policies and activities upon the environment.

Fortunately the Endangered Species Act commenced in 1973, though it has not stopped the extinction of many species in the U.S., and the Clean Water Act passed in 1972. Both these Acts have had politicians and industrialists continuously attempt to undermine them since their inception. It's not enough for them that under the Endangered Species Act a landowner is not required to do anything for the long-term survival of an endangered species. "In fact, it may be possible for a landowner to rid himself of an endangered species 'problem' by literally doing nothing and waiting until the habitat is no longer suitable for the species in question."[1] But waiting is not part of the DI. Worse, the Clean Water Act was fully exempted in regard to fracking for natural gas under VP Dick Cheney's leadership in 2001. Additionally, all of these environmental laws, and more, have not diverted the Forest Service, part of the Department of Agriculture (USDA), from its primary focus: prioritizing the forest for its timber only as a crop, resulting in the decimation of our forests and our companion species.

Conservation biology was not even thought of until 1978, but Jimmy Carter (and Congress) had the foresight during his presidency to double the National Wilderness Preservation System (NWPS), mostly in Alaska. However, the general rule is that "low-elevation, high-diversity sites are private lands; mid-elevation sites are commodity-production public lands; and large protected areas (such as designated wilderness) occupy the high-elevation, low-diversity sites. This biased pattern of habitat protection [is known as] the 'rock and ice' phenomenon...."[2] In other words, economic interests threw the public a bone, since natural resource extraction wasn't cost effective or plausible in the rock and ice areas anyway.

Conservation biology was a new DCI concept that came into play in the late 1980s and early 1990s when the controversy came up about protecting the endangered northern spotted owl or cutting down the rest of the old-growth forests in the Pacific Northwest, prompting in-depth research of forest ecosystems for the first time. It finally became clear to some in the DI that humans are not independent from nature as imagined, that there is not enough protected habitat for larger animals and top predators, and that the sustainability of ecosystems over the long-term must include the full spectrum of species, a term known as biodiversity, coined by E. O. Wilson and Peter Raven in 1988.

The realization concerning the importance of maintaining biodiversity brings us to the current day: that large preserves are needed to allow for the 3.4-billion-year continuation of evolution on its own terms, and that the natural world can't just be reduced to safe areas within the multitude of human activities without shutting down the process of evolution and causing a mass extinction of species.

In 1962, only 3 percent of the Earth's land was protected, mostly in the U.S.; today, less land is protected in wilderness areas than is covered over with pavement in the lower-forty-eight states of the U.S. Worldwide, 12 percent of the Earth is protected in various ways, but many of these areas are very ineffectively protected on paper only and subject to massive political corruption. Combined with less than one percent of the ocean in protection, only 3.8 percent of the Earth is supposedly safe from human activity. Considering how meaninglessly many of those areas are safeguarded from exploitation, only about one percent of the Earth is secure from man's abuse, and none of it is when one considers the climate change we're causing along with chemical and radioactive pollution.

By 2010, 68 percent of countries in the UN had amended their constitutions to protect their national environments, but not the U.S.; plus, what we really need is an Ecosystems Protection Act that aligns political boundaries with biomes or creates migratory wildlife corridors, not just having insufficient stationary parks that are non-adaptive to our anthropogenic changes.

In the 1990s, biologists, conservationists, environmentalists, and such began to face the actuality of a global mass extinction of species. This realization changed the perspective and focus of conservation to the well-being of whole continents for millennia rather than just getting by with the minimal amount of species possible in order to avoid extinctions by law while maximizing human development with compromised solutions that are ineffectual in the long-term. This inspired visionaries to come up with plans to create wildlife corridors such as Yellowstone to Yukon (Y2Y), which would include means for wildlife to migrate without human interruption with the full bio-inclusion of species all the way up to top predators[3], a much-needed paradigm shift in thinking, though as yet without legislative backing.

NORMALCY BIAS

PSYCHOLOGISTS HAVE A TERM they use for human behavior called the normalcy bias. It means we believe whatever situation we are born into is normal and is the way things have always been; any change we might be aware of is also considered normal. The normalcy bias acutely includes the cliché that we don't know what we don't know, greatly compounded if there is an unawareness of and an unconcern for history. This is particularly true as children when our brains are being programmed about how life is, since we have no sense of the past. "[T]he... human mind soon accommodates itself to circumstances... and regards as right whatever law or custom teaches is right. ...The beliefs to which persons have been habituated from childhood are... deemed [as] truths...."[1]

One of the primary functions of our brains is to develop and recognize patterns. As humans, we have become very specialized in this; furthermore, the job of our subconscious is to maintain the patterns we have developed in our lives. Even as conscious change occurs, our subconscious will usually succeed in snapping us back to old patterns to accomplish its perceived job of protection with familiarity, the normalcy bias. As we age, the patterns of our past experiences are overlaid onto our present and also influence our expectations of the future. We also have a penchant to believe that everything is okay, and we want to be happy, so we tend to focus on what is pleasurable even if it's unhealthy.

Our human proficiency at recognizing patterns also makes us vulnerable to seeing things differently than they really are, giving us a strong inclination to downplay any threats to our patterns. Aggregate social patterns develop that to various degrees coerce almost all people into buckling under social pressure; we still have an ancient, life-threatening, tribal need to fit in. This tendency to go along with collective denial makes us slow to adapt; we have a strong desire and unconscious drive to maintain the status quo even when we see there are problems. Our normalcy bias often prevents us from seeing any problems at all even if our lives depend on it, especially if not perceived as an immediate threat.

This dynamic is particularly true in the case of environmental change, since such change happens over longer spans of time than that of a single human life. A child born after a dam has been built loves the lake but has no idea what the previous river and the entire ecosystem was like, loves the new forest but has no idea what the logged-off ancient forest was like, loves the desert but has no idea it may not have even been a desert before cattle trampled and grazed it to ruins, loves the scenic vistas but has no real sense they were once filled with wildlife, and so on. This pattern of normalcy follows them into their adult lives: they accept the environmentally-degraded world as it is, incapable of seeing it as it was; are unable to admit the extent of the degeneration, not recognizing it as a problem; and will even attack or mock those that point out the problems in order to fit into society's patterns to avoid being controversial, shamed, or ostracized, even when such denial catches them up in an inescapable wave that propels them like lemmings over a cliff.

Whether they like what's going on or not, each generation of children will grow up thinking that things are basically the way they always have been; that the extinction of species is normal; that shortages of water is normal; that climate change is normal; that having as many kids as you can or want is normal; that suicide and mental illness and pharmaceutical drugs are normal; that endless debate and inaction is normal; that radical, positive change is impossible; that people who are killed or die trying to change things is normal; or that disunity, resignation, hopelessness, hedonism, confusion, addiction, and such is normal. "It is almost willful forgetting – the means by which our species, generation by generation, finds reasonableness amid the irrational destruction of... earth."[2] "They talk about the devastation of the rain forests, about deadly pollution that will be with us for thousands... of years, [and] about the disappearance of dozens of species of life every day…. And they seem perfectly calm… stupefy[ing] themselves with drugs or television at night… [unconscious] about the world they're leaving…."[3]

For all these reasons, catastrophic gradualism is another term used in place of the normalcy bias in regard to the ongoing mass extinction.

DOMESTICATION

IN RELATION TO WILD animals, humankind has always been vulnerable, so we've had to outsmart them and gang up against them. Later, we began to domesticate whatever plants and animals we most easily could, which also included killing all creatures suspect of preying upon this newly controlled food supply.

Conquest has always been about the food. The rise to cultural dominance of the world by Western Civilization started in Mesopotamia with the domestication of whatever animals and edible plants could be tamed. With the ability to create and secure its own food sources, the DC sprouted, and the human population began to grow, reaching the HEE within 10,000 years. The conquest of nature began with agriculture, and such domestication was also complicit with slavery, the taming of humans.

Originating from the power and wealth of food, wars have been going on in the Middle East ever since, long involving other Mediterranean countries and beyond. Trade routes of food developed among neighboring city-states and had to be protected from thieves. Conflicts were ultimately decided by food wealth and evolved into conquests of war. Alexander the Great, a Macedonian Greek, took over the known civilized world, defeating the agricultural Persian Empire of Darius III and others. Egypt, well-endowed agriculturally, then embraced Alexander and gave their empire over to him without a fight, grateful to have Darius gone. The world's greatest agricultural empires merged, expanded, and became more systemized and established. Much stronger, more secure trade routes were created.

The Greek cultural dominance was followed by the Roman Empire, which imported great amounts of grain from Mesopotamia and elsewhere to literally feed its strength. Western Civilization was firmly established, and it has remained the dominant world power in every way to this day, developing most in the USA, Europe, and Israel. Although guns were invented in China, almost everything else we use in our modern world is what the DC has invented; they did not come from Far East Asia, nor Africa, nor Russia, nor India, nor South or Central America, nor from Australia – inventions such as combustible engines, airplanes, sawmills, cars, diesel-powered ocean liners, telephones, plastic, rubber, computers, cell phones, conversion of electricity, wheels, oil drilling, washing machines, farm machinery, dishwashers, chemical fertilizers, refrigerators, pesticides, nuclear power plants, GMO's, vaccines, atomic bombs, the printing press, radio, fractional banking, television, microscopes, telescopes, and thousands more.

Creating order out of the natural world is what humans do, whether it's irrigation, the building of dams, creating homes, villages, and farms, or even braiding hair. With this need to create order, which enhances our security, comes domestication and increasing levels of manipulation and artificiality.

The developing culture of civilization got so good at manipulating life in its favor that they thought they were equal to God, at least created in "his" image, and definitely separated and superior from the rest of life on Earth, which were full of beasts forever stuck in their unchanging roles. The ideological gap widened when it was put into writing. Some in our species even named us the wise ones, the sapient ones (*Homo sapiens*), and officially sanctioned, in their self-generated sacred texts, that the world was made solely for them and given to them by God to do with as they wished. How convenient; the DC had a narcissistic meme to begin with.

By the time the Europeans were able to explore the world, the ideology of conquering all life on the planet, including other humans they saw as not much above beast status, was an unquestionably entrenched ideological goal, a self-serving purpose of superiority and greed born of cultural megalomania. The DCI came to the Americas, conquered all, and proceeded to reduce the entire world to purely that of human use – economic use - even now to the point of leveling mountains, fracking the Earth, and changing its very atmosphere.

Our self-serving domestication and manipulation of the world has brought us to the point of our ongoing anthropogenic mass extinction of species. Even domestic species are disappearing at an alarming rate; hundreds of apple, pear, potato, goat, cattle, and chicken species are now extinct, just to name a few, and many of the ones that are left, whether plant or animal, have to be artificially inseminated, often need human midwifery, are dependent on humans for their food and fertilizer, need our protection, and are largely incapable of surviving without us.

Already by 2000 BP, domesticated, quadruped meat eaten by the known world was down to just sheep, goats, pigs, and cows. The three foremost grain crops of our modern world amount to just wheat, rice, and corn for 80 percent of the world's people. Our consumption of birds is now reduced to just chickens, turkeys, ducks, and geese. In the past hundred years, the population of these domestic animals has multiplied much greater than our own, and just in the United States, we slaughter them for our consumption at the rate of at least 250 animals per second, which is "nearly 10 billion chickens and turkeys, 93 million pigs, 37 million cattle, 24 million ducks, 2 million calves," and a combination of 6 million horses, goats, and sheep annually as of 2002.[1]

We continue to kill and usurp any other animals we perceive as a threat to our domestic animal supply: prairie dogs, raptors, coyotes, bears, cougars, wolves, competitive grazers, and many more, further diminishing the unrestrained rhythm of evolution and any illusion of the wild. For example, in the Bighorn Mountains, Wyoming, the Bighorn Sheep for which the range was named are gone except for a few individuals with tracking devices around their necks, their habitat taken over and grazed by more than 50,000 domestic sheep[2], sheep well-protected with traps, guns, and poisons.

The oceans have also been tamed. Aquaculturists call it "the close of the life cycle" when they have fully domesticated an aquatic species. With the success of sea bass and salmon farms, the domestication of other valuable fish, such as Bluefin tuna, sturgeon, tilapia, and grouper, has taken place; they're neither born, bred, nor die naturally, living managed lives in total confinement. We also protect our oceanic food supplies as we do our domestic ungulates, formerly killing tens of thousands of bald eagles in the U.S. to supposedly protect salmon species, shooting and beating harbor seals and any other species seen as a threat to fisheries, the ongoing butchering of dolphins by Japanese tuna fishers in the Taiji Cove[3], the complete wasting of sharks just for their fins[4], and so much more.

The most domesticated species of all, however, is *Homo sapiens*; we'll even let the human-caused ongoing mass extinction happen without a whimper for fear of ruining our own individual lives.

FRESH WATER

ONE DROP OF WATER contains about 3 sextillion (3^{21}) water molecules; placed side-by-side, 100 million of those molecules would span about one centimeter.[1]

Our salty oceans contain 97.2 percent of our world's water, leaving just 2.5 percent in the form of fresh water, most of which exists in the form of glaciers, primarily in Greenland and Antarctica. A comparatively minute amount of water is suspended in our atmosphere as vapor. The amount of usable fresh water adds up to less than half of one percent of all water on Earth; fourteen percent of that is in the form of underground aquifers, the rest is in lakes and rivers.

Lake Baikal, in southeastern Russia, is the oldest, deepest, and largest lake in the world, holding one-fifth of our world's fresh water; the Amazon basin and America's Great Lakes each hold another one-fifth for a total of 60 percent of the world's fresh water in just these three areas.

Most of the rivers of the world have been tamed with dams; for instance, only three percent of the 1,243-mile-long "mighty Columbia" of Canada and Washington State is free-flowing (37 miles). The Colorado, Nile, Yellow, Ganges, and Indus rivers don't even reach their deltas for much of the year, if at all, which has radically altered those ecosystems, depriving them of their vast wetlands that supported enormous amounts of biodiversity. No longer is the Nile Delta replenished with silt and its nutrients because of the Aswan High Dam, built in 1970. The eastern Mediterranean is saltier and less fertile as a result, the sardine fishery destroyed, and the sea has since moved inland by over 1.25 miles, erasing five of the Nile Delta's seven ancient distributaries. "There are an estimated 800,000 dams on the planet and 40,000 large dams – an incredible 20,000 in China alone."[2] Dams control 60 percent of the world's rivers, and 75,000 dams of all sizes control nearly 95 percent of U.S. river miles. As small and old as they may be, there are as many as 5,000 dams in Connecticut alone.

Though fresh water covers less than one percent of Earth's surface, about one-third of vertebrate species are totally dependent upon fresh-water habitats.[3] The number of endangered fish species in North America is up to eight times more than either birds or mammals.[4] More fresh-water species are endangered and lost to extinction than in any other type of ecosystem; this is shocking when one compares this to the crisis of the world's coral.

Dams destroy ecosystems. For instance, the Glen Canyon dam in the Colorado River is trapping silt behind the dam, filtering it out of the river, plus making the river much colder by releasing the water from Lake Powell from down to a depth of 230 feet. This combination, along with unnatural ebbs and flows in the river made by human demands elsewhere, is pushing species native to the Grand Canyon to extinction in direct violation of the Endangered Species Act.[5] The peril of riparian systems in America's arid west has dire consequences to 75 percent of its native vertebrates. If not for the radical effect on river ecosystems from dams and the toxic pollution from farming chemicals, intense grazing, and the overdevelopment dams enable, the ongoing extinction of freshwater species would not be so extremely exacerbated.

Since 2006, China's rivers became so polluted that "80 percent of the [31,000 miles] of major channels could no longer support fish of any kind."[6] That same year, the Yangtze River dolphin – "the goddess of the river" - was declared *extinct*, with many more species preceding and following. The giant Yangtze runs red with pollution into our seas; hundreds of dead pigs from industrial farms occasionally float down others.

About 70 percent of global fresh water is used for agriculture. In the U.S. that number is 80 percent, 50 percent of which goes just to grow cattle and their feed. Just 20 percent of fresh water goes to cities, and out of that, billions of gallons go to growing grass on golf courses, flushing toilets, washing cars, and growing plants that often belong to completely different climates. As an example, Arizona extracts twice as much groundwater from underground aquifers as can be replenished; with new communities continually being developed, the water table has dropped more than 225 feet since 1950. This has caused more sinkholes and large fissures than anywhere else in the world, making homes and neighborhoods entirely disappear. Likewise, the enormous Oglalla aquifer which irrigates "the breadbasket of the world" in the Great Plains of America is expected to last only until about 2050. The Oglalla aquifer was filled by melting ice at the end of the last ice age and is not replenished with rain.

The water table also falls with the loss of trees. Trees are reservoirs of water, absorbing and regulating runoff, which greatly help to replenish the groundwater and stop erosion. Without trees the water table drops, and the soil, the water table, and even the rivers can dry up, such as the Rio Grande. With no evaporation, the surrounding environment receives less precipitation, continually worsening the situation, and overall temperatures rise. ("Impacts of deforestation on local climate: 20.3% less precipitation, 58.7% less soil moisture, 2.4% higher temperature."[7]) How much drier is the Earth, now that we have cut down 75 percent of the world's primordial forests?

As the human population continues to increase exponentially, and the depletion of fresh water reserves becomes a critical issue, some think that desalinization of the oceans for fresh water will be the answer that rescues humanity. However, desalinization plants are extremely costly and need to collect, pump, and then filter water through miniscule sieves at about 900 pounds pressure per square inch, which entails using massive amounts of energy.[8] It is not cost effective and does not produce enough fresh water to even meet supplemental demand.

The amount of cultural denial prevents necessary changes from happening to avoid the oncoming water crisis. The DC also completely ignores how much more a water crisis will affect animals even more than people. People still deny that climate change is happening, or believe the humanists will invent or discover something spectacular to solve the issue, or that nature will reverse any current droughts just in time.

The issues of the upper and lower basins of the Colorado River in the American southwest are a great example. Utah, Arizona, and

Nevada are the most arid states in the U.S.; California and Colorado are thirteenth and fourteenth. These states are also among the fastest growing, and it is primarily the gorgeous waters of Lake Powell from Glen Canyon Dam and Lake Mead from Hoover Dam that meet a great amount of the needs of these states. When plans were being made to control the river for electricity, recreation, basic human needs, and cattle it was the wettest the climate had been in this area for 450 years, as later discovered by tree-ring studies. People still haven't realized there may be a limit: the per capita usage in St. George, Utah, is 335 gallons per day, one of the highest water consumption rates in the country.[9] In greater Phoenix, in the Sonoran desert, it's 1,000 gallons – but at least they have the highest-spewing water fountain in the world.… In November 2007, in Mesa, AZ (greater Phoenix), voters overwhelmingly approved the construction of yet another water park to provide the desert community with surfing, snorkeling, scuba-diving, and kayaking, and instead of using recycled water, the park draws upon the underground aquifer.[10] Arizona also uses these waters to grow an abundance of cotton, alfalfa, and hay in the desert, crops very high in water usage, not to mention dozens of golf courses the area is famous for. The DIC has no true concerns for the future.

According to studies, "climate change could cut the flow of the Colorado River… by a crippling 60 percent."[11] When Lake Powell falls below its altitude level of 3,470 feet, water will not be able to generate any power. If policies are not changed in the way the Colorado River system is operated, "there is a 50 percent chance that both Lake Powell and Lake Mead will… reach dead pool by 2021."[12] This means the Colorado River would not even flow through the Grand Canyon, Las Vegas would not have any power or water, nor would many crops in Utah, Nevada, Arizona, and California have water.

Lake Mead needs only to fall below its altitude level of 1,050 feet for the first straw to shut off, causing a loss of 40 percent of current electrical generation and water flow, and 950 feet for the second straw, taking out the remaining 60 percent. As of this writing, work is far behind schedule to drill a third straw at 860 feet, officials are scrambling to build a multi-billion dollar pipeline system to take fresh water from northeastern Nevada to supply Las Vegas, and the levels of Lake Mead and Lake Powell are at 1,076 and 3,590, respectively (5/1/16). [13,14] That's just 26 feet to go for Lake Mead (rather than 170 feet when full) and 120 feet for Lake Powell (rather than 230 feet when full), and the levels drop faster as the flooded canyons narrow with depth. These measurements are expected to go lower even after this year's (anemic) annual spring snowmelt from the Colorado Rocky Mountains.

Incredibly, nothing has been done to change current water-use policies, no, that's too unthinkable. The DC would not even consider adapting lifestyles and businesses to fit nature; the DC would rather make its desperate attempts to maintain its very flawed paradigm. After all, bigger houses, more golf courses, and more water parks must be built; more hay, feed corn, cotton, and alfalfa need to be grown; and the lights and fountains of Las Vegas must not fail. Let nature control us? That's totally against the vainglorious DIC.

We'll keep hearing that more and more issues such as these are too big to fail, but the anthropogenic ongoing mass extinction – the biggest, most tragic failure in the 3.4-billion-year history of life on our planet – is not even recognized as a legitimate concern. The DIC will save its economy by bailing out its banks, but it hasn't even realized or considered that preventing species from extinction actually *is* incomprehensibly too big to fail. The DC can't even comprehend that it has been wrong for thousands of years; it would rather go down defending itself in violent denial and righteousness.

"[T]he world's wet places are being emptied. By some estimates, half of them have gone already. ...The Middle East ran out of water some years ago. It is the first major region to do so in the history of the world." [15] It's no coincidence that's where the DC originated.

We are the most narcissistic species to ever inhabit the Earth.

ARTIFICIALITY

"Mankind has gone very far into an artificial world of his own creation... and farther... into more experiments for the destruction of himself and the world." Rachael Carson[1]

IT IS IN OUR nature to control our environment to produce safety and comfort for ourselves. In the process of doing so, we have affected every lifeform on Earth in every part of the world. We have changed the chemistry of our atmosphere and our oceans and have begun the melting of ice sheets, affecting in these and other ways the most remote wilderness areas to everything else on down. In our attempt to dominate nature, we've progressed further and further into the creation of a manmade world. Prioritizing economic and industrial development, our Earth is simply considered a commodity, and we manipulate it with total disregard. By artificial means, we poorly attempt to restore or replicate what we have destroyed, only when we have to, and increasingly too late, especially compared to exponential times.

We have artificial grass, foods, plants, fish, and birds; fake hair, suntans, fingernails, and breasts; pseudo conversations with machines; simulated environments; and a countless number of other. Everywhere most people go, they are inundated with imitation, so much so that it has become normal and unrecognized. We've replaced wild forests with monoculture tree farms and wild prairies with fence-to-fence GMO crops. We are developing robotic bees to replace the sharp decline of natural bees. Rather than choosing to harness natural solar energy, we create artificial atoms through nuclear fission to produce electricity.

Simulated entertainment "exposes children to behavior that man spent centuries protecting them from." – Carrie Fisher.[2] We even have corporations deemed as real persons with all the same rights, resulting in fabricated persons dictating to all governments and making greater net profits than the GNP of many entire countries. Parts of China have simulated sunsets and sunrises on gigantic video screens in public places because the air pollution is so thick that people never see the sun, the horizon, or sometimes even the buildings across the street. Singapore has a mock tropical jungle, and the world has thousands of "urban jungles" full of simulated reality.

We are quickly moving into transhumanism where we will have artificial intelligence (AI) within our bodies. Our minds and bodies will be computer-enhanced; we'll become bionic humans, and millions even look forward to being these pseudo humans. We'll have nanobot (nano-robot) blood cells circulating all through us, including our brains. Some see AI as the ultimate triumph that will solve all of our problems; others anticipate it will make those who can afford it attain some kind of synthetic immortality.

Many drool over the next digital toy to distract themselves with faux entertainment to fill the psychological void they don't even know they have, not realizing how the loss of nature has caused an emptiness and lack of life within them, missing what they've never had the chance, desire, or will to even discover - *because they don't care.* To them, the natural world conjures up thoughts and feelings of danger, discomfort, disgust, demons, and disease. At most, they might have a vicarious, imitation nature experience from some digital device when they're bored. Many even hate nature, something that must be endured at best. Nature has never been their friend, let alone their lover. Treating nature as something inanimate and unfeeling to rise above or destroy, they remain completely insensitive to it.

The natural world is departing swiftly; even the Earth systems that have kept our planet in balance are being thrown out of whack. Like Humpty Dumpty, once parts of the biota of Earth go extinct, scientists will never be able to put it all back together again. The planet will be too humanized anyway; the former lives of nature won't even have a place to exist except with their DNA in tiny vials placed within liquid nitrogen stored within a vault deep inside a mountain.

Humans now decide which species will live or die, even with the Endangered Species Act; we have taken over the role of natural selection and have caused the end of the wild. We've bought into our own lies so much and are so out of touch that "we drive ten-mile-per-gallon armored transports to move groceries home,"[3] have recreational water parks and million-gallon-a-day golf courses in the desert where we also grow heavily water-dependent hay and cotton.

We have grown up enough as a species to have separated from our mother, nature, the Earth, yet we are still like babies trashing our environment, thinking some parental higher power will clean up after us, thinking that life is all about us and our needs, that everything centers around us, and demanding to have what we want when we want it. We have not grown up and matured with responsibility as a species at all, acting as if humans are all that matters.

We are imposing our manmade world over the broken natural world we caused to break, and in our arrogance we think we have the solution, at least pending, to every problem. Our species will live in a world completely concocted by humans or will try to survive in a desertified world; either way, our Earth will be far out of balance and devoid of at least half its species. The DC sees it as progress, replacing the wild and problematic world of nature with a world "improved" by technology.

The artificial world is taking over as the other vanishes into what is becoming an unknown antiquity. We're leaving our verdant, gorgeous Earth behind with no way back; we are creating a new Earth. The Earth is forever changed. The bread crumbs marking our trail are gone. "[W]ithout wilderness, the planet will sink further into biological poverty, and humanity's communion with its roots will be lost forever."[4] We're in totally new territory; we've even changed the weather patterns for the entire world. Even if humans all disappeared today, it would take at least 100,000 years for the Earth to purge itself of our manmade creations.

All creatures on Earth are dependent upon what we do; many critically endangered species need our endless attention, with Radio Frequency Identification (RFID) tags around their necks, on their ears, under their skin, unable to survive without our constant management. Many need "artificial insemination, test-tube fertilization, electro-ejaculation, or... [s]emen... stored away in frozen test tubes. The fertilized embryos of endangered gaurs are being implanted in Holstein cows. Horses are giving birth to zebras."[5] The amount of stuffed, extinct creatures on display at prestigious museums continues to grow. Ecosystems absent of their apex predators become unnatural. Oceans become deserted as we instead raise fish on farms, domesticating any we want.

Rather than addressing the root cause of our problems, the DC proposes ludicrous, engineered schemes: sending gigantic mirrors into space that would reflect the sun to counteract global warming, fertilizing the oceans with iron and creating new GMOs to promote more carbon absorption, and even modifying the ocean currents with giant barriers to distribute heat around the globe....[6]

Rural areas are cut up and planted in manmade squares; animals are injected with hormones, antibiotics, and semen to artificially keep them alive; farms and ranches subsist on synthetic chemicals; and rural economies are propped up with counterfeit government subsidies. Yet the cities are even further removed from reality with stars blocked out at night, sirens blasting and manufactured lights blazing, and nothing but a few domestic plants and animals and weeds to substitute for nature.

All of this continues to separate us from the natural world, a world the DIC still thinks it can conquer, dominate, and control to its own will in its quest to rival "God," convinced it even knows what God is or wants, conceitedly relating itself to some self-defined version of "Him."

ECONOMICS

T HE BOTTOM LINE IS our economic system is one of the primary causes of our ongoing mass extinction of species, our Human-caused Ongoing Mass Extinction, our HOME. The economy has become the all-powerful secular god, and it is global in scale since "free market ideology [capitalism] has been embraced around the world with near-religious fervor"[1]; thus, the economics of the DC rules the world. Even former communist countries have developed their own brand of capitalism. The health of each country's economy has the highest priority, which the DC believes requires continual growth; otherwise, it's believed the economy will recede, the country's international power will be lost, and its people and way of life will suffer.

DC economics "assumes an imaginary world divorced from reality... [and] to question this doctrine has become heresy...."[2] Economics mimics natural law (more rabbits equals more coyotes, like more corn equals more humans), but humans have developed natural law into a highly sophisticated system full of conflicting economic theories simultaneously applied, going back to the 1700s with Adam Smith, then John Keynes, Milton Friedman, and others in the 1900s. Within these theories are complex graphs and charts and terms such as opportunity costs, comparative advantage, economies of scale, externalities, and the more familiar supply and demand. Economic laws have become the humanized version of natural law, highly manipulated with taxes, subsidies, tariffs, price ceilings, price floors, and more.

Historically, economics (which drives politics) has superseded the care for the environment every time. It wasn't even until the beginning of *this century* that economics and ecology finally came together; the new field is called ecological economics. This came into being after enough scientists and scholars started to realize the value of natural services, now that those services are being altered and jeopardized and can't be taken for granted; now that all the other reasoning to save the environment hasn't worked; now that we realize how much it would cost to replace these essential services even if we could, and if we can't, how all systems would crash; and now that ecologists know the only thing the DC *might* listen to is economics because, to the DC, money is the only measurement of value. Natural services "provided freely by nature are treated as valueless, until scarcity and privatization render them marketable."[3] To the DC, all of nature is a commodity, and all that can't be used is worthless and unimportant.

Opportunity costs are the possibilities lost by having to make choices. This applies both in business and personally. In a monogamous society, one loses the opportunity to marry one person by marrying another, or if one has a two-week vacation and wants to go to Italy or Hawaii, one opportunity is lost by choosing the other. Businesses with limited funds do this all the time with their investment choices; likewise, conservationists do the same by having to choose which habitats to protect and restore, unable to do all, like performing ecological triage as in the case of West Africa.

Comparative advantage is the ability of a nation to produce raw materials or products at a lower opportunity cost than another nation, so it is able to focus all its energy on being really good at the thing it has chosen to produce rather than also producing what other countries can produce better. Also relative is absolute advantage: the ability of a political area to produce raw materials or products using fewer resources than another political area (i.e. a country having a natural proclivity to produce a certain material, such as Saudi Arabia with its oil). Comparative advantage comes into play when two or more countries can all produce the same things, such as automobiles, rice, and beef, or weaponry, pesticides, and natural gas, but they all choose their best opportunity costs to gain comparable advantage as a specialty in something, adding to whatever absolute advantage they may also have. There's no reason for other countries to compete with them at certain things; it is far more advantageous, especially in a global market, to produce the things you're better at than anyone else. The comparative advantage becomes a country's specialty, choosing that over its other opportunities. Such specialization leads to the exploitation of local resources, resulting in the loss of diversity in both the landscape and in species that are essential for the proper functioning of ecosystems, such as Canada liquidating its boreal forests (an absolute advantage) for paper pulp.

After WWII, there were two superpowers, the U.S. and the USSR, both competing for military advantage to be able to spread their version of the DC. The costs of the Chernobyl accident in 1989 broke the Soviet's ability to compete, it was the year the Berlin wall came down, and the empire of the USSR shrank to become the present form of Russia. As a result, the U.S. became the world's only superpower and developed a huge comparative advantage in producing high-tech armaments of every kind. More than anything else, it is this one industry that maintains the U.S. economic superiority over any other country. In fact, the U.S. military budget exceeds that of all other countries combined in the world when factoring in the funding of ongoing wars with supplementary bills outside of the federal budget; without this extra funding, the regular U.S. military budget is still more than the next twelve countries' highest military budgets combined, including China and Russia.

As the epitome of the 5,000-to-10,000-year development of the DIC, it is the production of weaponry and the military dominance of America that maintains the order of the world and its paradigm, economically and ideologically. The U.S. is the number-one supplier of warfare in the world, amounting to "nearly half of all the arms sales in the world."[4]; without being so, the U.S. would lose its version of control over the current paradigm due to the loss of economic power. Our consumer wealth also props up all the other economies of the world; that's why China and Japan keep lending the U.S. money, creating a positive feedback loop of its own. "Massive defense contracts... have now become the most important part of the American economy. These programs produce such a prosperity that it was necessary to extend the war economy of the last [four] decades to prevent wholesale collapse of the social structure."[5]

Other countries are trapped into the same system because U.S.

trade, born of massive consumerism, keeps their economies alive, too. And as people in an increasingly humanized world desire and accept consumerism as humans' purpose in life, even more see the world as nothing but a vast supermarket. In 2011, the illegal markets of wildlife smuggling alone generated $20 billion per year, making it the third-largest source of black market income after drugs and guns. Next were illegal fishing, $16.5 billion; illegal logging, $15 billion; and illegal garbage/hazardous waste trafficking (including radioactive), $11 billion.[6]

Economies of scale is a term used when a company produces output of product at the least amount of cost per unit. In a competitive market, all companies try to achieve this to give them a competitive advantage. For example, a factory producing shirts needs a certain amount of space and machinery to produce a thousand units per day, but they could make ten thousand shirts per day with the same equipment (and perhaps the same labor), lowering the cost per unit, which maximizes profit. This applies to anything from producing wool hats or cat food to forestry products or barrels of oil.

Especially since the 1980s, companies have been looking how to streamline operations in any way they can in order to exist in a very competitive market. The economic principle of economies of scale has given rise to such machines as feller-bunchers that can harvest entire swaths of trees faster than ever imagined, replacing thousands of forestry workers, but feller-bunchers have to be operated at overtime pace to make the cost of the machines worthwhile. This means forests are being destroyed at record speeds because of manmade economic necessities, which are mere human mental concepts, like the insanity of Maxxam Lumber mowing down ancient redwood forests to pay for junk bonds two decades ago. The difference now is that directly exploiting nature is happening as fast as possible in every industry around the world. It is also happening indirectly, as in the case of the Asian restaurant industry still serving endangered species such as Bluefin tuna and sea turtles or shark fin soup.

Supply and demand is generally understood except that, when a product successfully reaches a point of high demand, more producers of the product will join the market because that's where the opportunity is. Over-production then drives down the price, but then government policies kick in to artificially support the market. These policies take the form of subsidies, a form of corporate welfare, or tariffs, which help curtail the competition from foreign countries of similar products. Both policies help prop up comparative advantage to keep a country's economy artificially healthy; subsequently, the reality is the free market is an accepted illusion.

An example of this is the U.S. corn market. When a giant increase in demand was created by the government's decision to make corn ethanol, many producers chose to grow corn because it became a better opportunity, but then the resulting glut of corn drove the prices down. Farmers then started receiving giant government subsidies to continue to grow more corn, a very toxic crop to the environment, full of negative externalities the public pays for while the industry massively profits. Every industry that involves natural resource extraction follows this example: coal, oil, natural gas, forestry, fishing, nuclear, and many more. All the market manipulations make these products profitable *and* affordable, but the costs of the negative externalities all are at the expense of our planet and all its lifeforms, pushing too many into the threat of extinction.

Externalities, usually referred to in a negative sense, are the costs of doing business that businesses are not held accountable for that are passed on to the world at large. Examples are everywhere such as slash burns after logging operations that generate massive air pollution, or nuclear plants creating deadly radiation, or bottom trawlers destroying the structures of benthic life, or frackers polluting pristine underground aquifers with highly toxic chemicals and adding methane to the atmosphere. Externalities exemplify "the tragedy of the commons," when a few benefit from trashing the environment through natural resource exploitation at the expense of the many. Large corporations have become very adept at privatizing profits and socializing losses such as the hundreds of billions of dollars in the bank bailouts of 2008. The banks also bore no liability for their own errors, and no significant laws or policies were even changed afterwards.

Externalities are not factored into the costs of production; they are not reflected in the price of the final product. The price of gasoline, for example, does not include all the poisonous fumes that are expelled from gas-powered vehicles into the air, or the price of wars to protect oil reserves, or the interest payments on the borrowed money for those wars, or the benefits paid to the veterans of those wars, or the costs of oil spills such as the Exxon Valdez in the Gulf of Alaska or the British Petroleum oil leak in the Gulf of Mexico and on and on. No one could afford to drive if all those externalities were factored into the end price. Businesses could not continue if they had to absorb the costs of such externalities; instead, the environment absorbs them and species increasingly go extinct.

To show how warped the mindset has become by economic principles, investors in the Japanese whaling industry realized that maximizing the exploitation of whales would enable them to reinvest quick profits into new growth industries to increase their original return on investment. By hastily "liquidating" whales even to the point of extinction, profits would be much greater than if to preserve the species by taking them at a sustainable rate.[7]

"There is something fundamentally wrong with treating the planet as if it were a business in liquidation."[8] The DC's economic rationality has evolved into determining how natural resources are to be exploited; "[i]ndustrial development has confined the natural environment to the role of being a supplier of resources [and] a repository for waste…."[9] Any wildlands leftover for recreation are much of what is known as "rocks and ice" because it's where industries of natural resource extraction were unable to physically go, or because it wasn't economically feasible – an easy opportunity cost to waive - so most of it consists of amazing scenic views inside parks where more and more species are surrounded by industry and becoming more scarce.

The European, centuries-long colonialism that began 500 years ago changed the economics of the world, a change that still profoundly affects the world today. Colonialism shifted the wealth of continents to Europe, which then spread to take root in the U.S., the land of opportunity - for Europe. Europe confiscated massive amounts of gold and silver from North, Central, and South America; it took Europe's economy to a whole new level of affluence and power. Europe did the same with its colonies in Asia and Africa.

Before WWII, this wealth allowed Europe, North America (above Mexico), and Australia to become developed countries (the OCCs) while the rest of the world, its riches of natural resources pilfered, were left in poverty, most still in a state of recovery today from the theft. They're known as "underdeveloped" countries (the OFCs). The countries of the tropics were the most exploited because that's where the most of everything above ground grows: food, spices, mahogany, rubber, ivory, teak, coffee, tea, sugar, and a cornucopia of other goods besides what could be mined. The OFCs became powerless, impoverished, and out of control. Now they're over-populated, which causes an over-supply for the demand in labor, so they still suffer from neo-colonialism (exploitation via globalism) in the form of very cheap labor, slave labor. However, more damage is done to the environment as desperate people look for any way to make money by any means, and endangered species have a higher value than anything....

The rich countries are still being served and protect themselves with ingenious economic ploys to always stay ahead. One of the major ways are the debt-for-nature swaps the International Monetary Fund (IMF), World Bank (both created in the U.S.), and World Trade Organization (WTO, with its many associated trade agreements such as NAFTA) make with the OFCs where their natural resources are taken as collateral for the debt obligations they can't pay. Colonialism first stole as much of the OFC's natural resources as possible, and now neo-colonialism requires ownership of the remainder in place of the defaulted debt owed them. The OFCs had to borrow the money from the nations that had it so they could invest in trying to get out of their poverty, which was caused by the OCCs to begin with.

The point is all of these macroeconomic issues are a major cause of the mass extinction of species that is occurring; the DNA of Earth is paying for all of our species' artifices.

One of these major artifices is fractional banking. This allows banks to create enormous amounts of money out of thin air by loaning out a large fraction of the deposits they receive. The deposits they receive on the loan payments further allow them to create more loans to make more money with to create more loans.... The loan money isn't tangible money, but the payments are. Then the Federal Reserve bank, which is a private corporation, backs all the fractional banks by simply printing as much fictional-but-physical money as needed to keep the illusion going. This money is backed by nothing except "the government's ability to pay" (to pay the debt paper money represents, which it can't). This is a highly manipulated scheme that maintains the ones in power. So the reality is our monetary system is even more of an accepted illusion than that of the so-called free market, and it maintains the entire economic structure of the world.

> We are experiencing accelerating social and environmental disintegration in nearly every country of the world — as revealed by a rise in poverty, unemployment, inequality, violent crime, failing families, and environmental deterioration. These problems stem in part from a fivefold increase in economic output since 1950 that has pushed human demands on the ecosystem beyond what the planet is capable of sustaining. (On average, we have added more to total global output *in each of the past four decades* than was added from the moment the first cave dweller carved out a stone axe up to the middle of the [twentieth] century). The continued quest for economic growth as the organizing principle of public policy is accelerating the breakdown of the ecosystem's regenerative capacities and the social fabric that sustains human community…. This has concentrated massive economic and political power in the hands of an elite few whose absolute share of the products of a declining pool of natural wealth continues to increase at a substantial rate — thus reassuring them that the system is working perfectly well.[10]

A second of the major artifices is the creation of derivatives, another manipulation of the illusion of money taken to the extreme; however, people make billions of dollars with them, although derivative losses were also a large factor in why the banks needed the bailouts of 2008. Derivatives are contracts whose value is derived from the price of something else, usually stocks, bonds, or commodities. Derivatives are simply high-stake bets, and the winnings are privatized, and the losses are socialized; the elite receive and society pays.

> The decisions of the financial system are increasingly being made by computers on the basis of esoteric mathematical formulas with the sole objective of replicating money as a pure abstraction. ...[Derivatives tie] up banking system funds in activities that are of questionable benefit to society when the credit needs of home buyers, farmers, and productive businesses go unmet. ...The reality that the [elite] are loath to acknowledge is that financial institutions once dedicated to mobilizing funds for productive investment have transmogrified into a predatory, risk-creating, speculation-driven, global financial system engaged in the unproductive extraction of wealth from taxpayers and the productive economy"[11] [Very prophetic words from 1995].

This is causing alienation away from the natural world as far as anything has gone in the human mind, enriching and empowering the ones who have the most influence.

A final economic truth of note is that people can only eat so much. In an economic paradigm that requires growth, this is not good for the food industry because it can only grow at the rate that people do. Given the paradigm we live in and that the food industry must survive in, is it any wonder that obesity is on the rise throughout the OCC world? Is it no wonder that our overpopulation problem is so strongly denied and that "more people equals more profit"? Industrial farms and the rest of the food industry are doing everything they can to push more food into every body, including adding addictive ingredients, and in its frenzied rate of production, many processes and ingredients are overlooked.

But at least it's good for the economy: health care adds very nicely to the GDP....

RHINOCEROSES

RHINO HORNS CONSIST OF pure keratin, the main protein our fingernails and toenails are made of. Nonetheless, this same protein is "worth five times more than its weight in gold in Hong Kong, Singapore, and the Mideast...."[1] Each horn's average weight is two kilos (4.4 pounds), so whether it works as a medicine is not the point; the business is too profitable for the illegal black market to go away. Rhino horn is used not as an aphrodisiac but as an unproven folk medicine for fever, rheumatism, gout, and most recently, cancer. Vietnam's demand for rhino horn is huge, as cancer is rampant in that country and growing by 25 percent per year, and very few treatment centers exist. Desperate people are therefore willing to try anything if they can afford it, even with the knowledge that rhinos are an endangered species. Rhino horns are also carved into objects used by the rich as status symbols – the lowest status possible – symbols of the most oblivious, uncaring, and bigheaded. Additionally, now that alcohol consumption is way up in Eastern cultures, many believe that rhino horn cures hangovers, so snorting it at parties is done by some. One could have a lucrative business powderizing the clippings of human fingernails if they were dishonest enough to say it was rhino horn....

Rhinoceroses are herbivorous pachyderms. As an umbrella species, their role in the ecosystem is the maintenance of the savannahs. They are selective mega-grazers that increase plant diversity, keeping the grasslands healthy for a wealth of other species such as buffalo, antelope, impala, zebra, gazelle, and many more, as well as birds and reptiles. More importantly, the mega-grazing of rhinos prevents super-fires from burning up the grasslands and beyond. The wildlife of the savannahs depends upon rhinoceroses for their food supply by stopping catastrophic fires from happening.

There were millions of rhinoceroses up until 1900. At that time there were a million black rhinos alone. Now, only one rhino species out of five, the African Southern White, is not listed as endangered. The only other white rhino subspecies, the African Northern White, purportedly has only *six* individuals left as of October 2014, with just one male capable of breeding. Black rhinos have suffered a 94 percent drop in population since the 1960s due mostly to poaching, and one of its four subspecies was declared *extinct* in Nov. 2013. There are believed to be fewer than one hundred Sumatran rhinos left in the world, and "the Javan rhino, which once ranged across most of southeast Asia, is now among the rarest animals on earth, with probably fewer than fifty individuals left, all in a single Javanese reserve."[2] Illegal hunting by organized crime and loss of habitat due to human encroachment are what is eliminating these ancient mammals. Often, horns are sawed off with parts of their skull while rhinos are still alive and then left to bleed to death in the shock of torturous pain.

Rhinos have been on Earth for 40 million years. In the time of just 2.6 ten-millionths (115 years) of their existence, rhinos are likely soon to be *extinct* because of humans. If 40 million years equaled 400 days, that's equivalent to the sick ideology of *Homo sapiens* having taken a mere 2.3 seconds out of thirteen months to eliminate rhinos from the Earth forever.

EVOLUTION OF *HOMO SAPIENS*

D NA BEGAN THE SEPARATION of chimpanzees from gorillas at about seven million BP. It took another 4 million years for the speciation from chimpanzees to manifest as *Australopithecus africanus*, an omnivore, and *Australopithecus robustus*, an herbivore, both of which faded away after an existence of 1.8 million years. It is from *A. africanus* that *Homo habilus* sprang, the tool users, becoming the first identified version of human rather than ape or between: this first human species began at 2.5 million BP, overlapping with the existence of *Australopithecus* for about 1.3 million years.

Homo habilus developed into *Homo ergaster,* also known as or confused with *Homo erectus*, the first hominid to migrate out of Africa at 1 million BP. (It's easier to call them *Homo e/e*). Fossils show that *Homo e/e* migrated to the Caucasus Mountains north of modern-day Iran and expanded into Asia. *Homo e/e* also expanded into Europe and evolved into *Homo neanderthalensis*. Additionally, *Homo e/e* remained in Africa and evolved into *Homo sapiens* by 195,000 BP before *H. sapiens* slowly ventured into the rest of the world.

Homo neanderthalensis, came into being at roughly 275,000 BP. As *Homo sapiens* started their expansion out of Africa, they eventually shared presence in Europe with Neanderthals, and Neanderthals became extinct at 30,000 BP after ten to twenty thousand years of contact with *Homo sapiens*. The Neanderthals were physically superior to *Homo sapiens*, but *Homo sapiens* eventually supplanted Neanderthals presumably due to superior language skills. Language may be what separated *Homo sapiens* from all other primate species, including human, by creating our ability to work in unison, to help one another improve. Neanderthals and *Homo sapiens* mixed together enough that today all humans have

1-4 percent Neanderthal DNA except for the *Homo sapiens* that stayed in Africa and never mixed. For the more than 200,000 thousand years that Neanderthals lived in what became Europe, they did so without any noticeable harm to their environment.

By 500,000 BP, the use of fire by the few existing species of humans had spread throughout Africa, Asia, and Europe. As it takes much more of the body's energy to digest raw food, the cooking of food, which no other lifeform can do, freed up energy that enabled the further evolution of the human brain, including the advancement of intelligence able to create language capabilities superior to any other lifeform. This was simultaneously fostered by the need of life-force self-willing DNA to make complicated language possible by evolving new physical attributes, as in the larynx, in *Homo sapiens*.

Rudimentary language probably first developed with group hunting, which involved the use of strategy. The use of fire further caused more language and brain development, as people gathered around for long periods of time to eat and stay warm. Our *Homo sapiens'* aptitude of complex language, a specialty unique to our species, has allowed us to pass on all of our acquired intelligence from one generation to another and continually build upon our abilities. Neanderthals were likely not as capable at passing on acquired knowledge and got bypassed by evolution, by natural selection. The inability to successfully pass knowledge on to each successive generation, due to the incapability to speak or understand sophisticated language, is the same reason other animal species stay basically the same throughout their roles on Earth. That is one fundamental reason why they are suffering from our HOME, following the same fate as the Neanderthals.

A few species of hominids had twice existed at the same time in various parts of the world, but circa 1.2 million BP, natural selection left only the *Homo habilis* lineage to continue, later becoming *Homo sapiens*. There was a time, at 100,000 BP, when no more than 10,000 adult *Homo sapiens* existed on our entire planet, but by 50,000 BP an evolutionary awakening occurred in *Homo sapiens* known as The Great Leap Forward. By 30,000 BP, Cro-Magnons (so named for the location in France where the remains were found and now referred to as Early Modern Humans [EMH]) emerged from the late Stone Age. EMH became the only human species left on Earth of all that had existed in the past 2.5 million years. A very adventurous gene eventually prompted modern humans to courageously begin venturing out to sea away from sight of land, which forever changed the course of history.[1]

Long before EMH, our ancestor hominids were not so much hunter-gatherers as they were opportunistic foragers that ate whatever came their way. Australian Aborigines call this way of living the Walkabout, which is likely what hominids did throughout the world.

When early humans progressed to hunting larger game, long-distance running had become an evolutionary specialty of our species and kept our ancestors alive for 200,000 years before spears or arrows were ever even invented. With the ability to stand upright, early humans were able to get much more air into their lungs. Early humans hunted in small bands that ran their prey down; while running, four-legged animals can only inhale when their legs are extended, while humans can inhale any time at any stride or pace. Humans also lost their body hair and developed superior sweat glands over their fur-covered prey.[2] A third trait was the ability to communicate, to strategize. These three traits allowed them to outlast their prey, prey that would overheat if not allowed to rest. The ease of the kill was greatly enhanced when early humans caught up to their target, a quarry too exhausted to fight or even stand.

Every species on Earth has at least one particular specialty of its own. From monkeys, DNA caused apes to lose their tails as they came down from the trees, and from apes, humans stood more erect and became long-distance runners as a means to survive and evolve into becoming proficient communicators.

The ancient Greeks were the first to define communication, using the word "communo": to share, to make known, to be understood, and to create common meaning. Aristotle wrote the first textbook on communication about 2,300 years ago; at that time, the form of public media was oratory. This did not change for over seventeen centuries until the invention of the printing press, causing the ability to communicate on a mass scale, forever altering the totalitarianism of monarchies and religions with the media of paper. It took nearly 500 more years before we understood electromagnetic frequencies and the use of electricity enough to invent radio; this began our exponential rise in our ability to communicate. Our various techniques of communicating wirelessly have since taken on many forms, completely changing how we communicate at every level and function of culture. From the media of vocal cords, to paper, to electricity and the harnessing of electromagnetic frequencies, to superconductor technology, to soon having phone devices that translate foreign languages in real time, our ability to communicate has grown from an orator broadcasting to a small group of people sharing the same language and culture to soon being able to communicate to anyone of various cultures on Earth without language barriers. Our ability to communicate in so many ways has been the greatest attribute of our species and continues to exponentially increase our capabilities, hopefully in time to profoundly cause an awakening in consciousness, an enlightenment towards how we relate to the rest of life on Earth and an intolerance in the completely unacceptable ways in which we have so far.

INSECTS

INSECTS ARE ARTHROPODS, WHICH originated in the sea. Though fossils show that some insects existed on land about 200 million years ago, they did not proliferate as we know them until about 150 million BP, preceding the first flowering plants, angiosperms, by 10 million years or so. There are about 900,000 known insect species[1], 40 percent of which are beetles.

Insects would survive fine without us, but we could not live without them for the natural services they provide.[2] We could say we're in the Age of Insects as easily as we are in the Age of Mammals. Insects are obvious pollinators, yet some of them are dying off rapidly: there are 97 percent less monarch butterflies since 2005, and honeybees are declining globally by as much as 70 percent in many areas. It is apparent that manmade chemicals, GMOs, and microwave technology are some of the combinatory causes of their demise. Another natural service of insects is in turning and rejuvenating soil more than any other organisms, including earthworms.[3]

Besides the estimated 10 quadrillion ants in the world[4], there are an almost equal amount of their termite relatives. Together, they nearly double the weight of the entire human population. With so much tropical rainforest having been laid to waste, the population of termites has exploded.

Thousands of books have been written on this fascinating and extremely diverse arthropod phylum within the Kingdom of Animalae, and however much we may try to kill, control, and manipulate arthropods, we also do so at great risk. As with bacteria, the majority of insects are beneficial to humans, and our technology and use of chemicals does not discern well amongst these species.

CORPORATIONS

"And by reason of their boundless pride... there shall be nothing remaining on the Earth or under the Earth or in the waters that shall not be pursued and molested or destroyed." Leonardo DaVinci[1]

WHEN EUROPE BEGAN COLONIZING North America in the sixteenth and seventeenth centuries, large merchant companies were set up by England and the Netherlands, giving birth to modern-day corporations. The monarchies granted charters to New World companies with the power to act as self-regulating entities. Such corporations had specific duties on behalf of the crown, which could abolish the charter at any time.[2] As the crown and its representative corporations became increasingly repressive, cause was given for the American Revolution.

After the Revolution, the chartered corporations were highly regulated, and the power to issue corporate charters was held by the individual states, allowing the citizenry to have a close watch over them. State legislatures maintained the right to revoke any charters that failed to serve the public interest, and the charters were limited to a fixed number of years, requiring renewal in order to avoid automatic dissolution.[3] Corporate members have since eroded and manipulated the system to assure their companies' own security, power, and autonomy.

In 1886, the U.S. Supreme Court ruled that private corporations had the same rights as persons under the U.S. Constitution *(Santa Clara County v. Southern Pacific Railroad)*; ludicrously, corporations became people by law. This makes no sense for a number of reasons. Death, for instance, happens to real people, but corporations can live and grow without limit and build power and wealth indefinitely. Examples of corporate longevity are: DuPont began in 1802; Standard Oil, Atlantic Richfield, Carnegie Steel, and Scott Paper incorporated in the 1870s; Exxon, Mobil, AT&T, Westinghouse, and Johnson & Johnson in the 1880s; Royal Dutch (Shell Oil), General Electric, and Dow Chemical in the 1890s; and Weyerhaeuser, U.S. Steel, Texaco, Ford Motor, and General Motors all by 1908.[4] This would make a real person alive and at their healthiest at up to 213 years old, born within months of the Louisiana Purchase.

Just fifteen years after Lincoln's Gettysburg address, President Hayes (1877 – 1881) stated: "This is a government of the people, by the people, and for the people no longer. It is a government of corporations, by corporations, and for corporations."[5] Those were the classic days of the robber barons, but today's robber barons we never hear about, ones such as the CEOs of Raytheon, Lockheed-Martin, GE, Boeing, and other high-tech armament suppliers, who each make tens of millions dollars of personal income a year, before investments. The modern-day robber barons aren't limited to corporate arms dealers: Monsanto, Syngenta, Novartis, Goldman-Sachs, Chase Bank, Mitsubishi, Sumitomo, and hundreds more from several countries are in every sector of the global economy.

A new world order was created in 1944 towards the end of WWII with the Bretton Woods summit in New Hampshire. It was there that U.S. and Britain led forty-four attending countries into an agreement to set up two banks operated from and controlled by the U.S.: the International Monetary Fund (IMF) and the World Bank. The U.S. dollar was established as the world's currency and the common denominator for all international exchange. The IMF was set up as a means to rebuild the infrastructure of countries destroyed by the war, and the World Bank was created to supply large loans to countries to grow; together they became the exporting financial institutions and debt collectors for large Northern-based corporations. U.S. corporations filled the demand for WWII reconstruction materials, and the U.S. gained control of many other countries' foreign policy as a form of collateral for World Bank loans. Post-war policies greatly enriched the U.S. economy and its corporations.

From that point, the natural development of communications, business organization, and technology gave rise to the global economy. In the 1980s, markets became unregulated; in 1999, the Depression-era Glass-Steagall Act was replaced, allowing investment banks to once again merge with commercial banking (derivative "investing," which is simply a competition between computers, then ran wild and led to the trillion dollar bank bailouts of 2008 and on); and by 2001, over two-thirds of the largest fifty economies of the world were corporations. Corporations were no longer bound by loyalty to any government in the world nor controlled by them. Multi-National Entities (MNEs) were free to "function without restraint and with accountability only to themselves."[6] The role of democratic government was reduced to the protection of private property[7], property which MNEs, as persons, have a "right" to exploit for all its worth - destruction of property (habitat) being the number one cause of species extinction.

The General Agreement on Tariffs and Trade (GATT) treaty, born on Oct. 30, 1947, was furthered by the U.S. Congress in April of 1994. NAFTA had been passed on Jan. 1, 1994, which "served to create and enforce a corporate bill of rights protecting the world's largest corporations against intrusion in their affairs by people, communities, and democratically elected governments."[8] The World Trade Organization (WTO) was then established a year later on January 1, 1995. These international treaties overrule national and local governments, even as stated in Article VI, paragraph II of the U.S. Constitution. By means of new free-trade agreements, previous political colonies have become economic colonies controlled largely by the MNEs through the World Bank and the IMF; this is referred to as *corporate colonialism*.[9] Free trade only ever meant that MNEs could "plan and organize the world's economic affairs to the benefit of their bottom line, without regard to public consequences" and avoid all the limitations of democracy.[10]

A new era has come upon us. Corporations have now come to the point of essentially owning or controlling the natural resources of the entire planet[11], a planet considered by MNEs to be full of no more than commodities to be exploited, including labor. This is venerated by the highest members of the DC as the latest manifestation of its long-standing ideology; from its outset, the DC has persistently separated itself from the sacredness and beauty of nature. "The rights of the natural world of living beings are at the mercy of the modern industrial corporation [MNE] as the ultimate expression of patriarchal dominance over the entire planetary process."[12] This originated with religious ideology and evolved into secular economic ideology, both based upon the heartless domination of people, nations, and nature itself.

The latest gains of corporate power over government and citizens involve the U.S. Supreme Court decisions of *Citizens United v. Federal Election Commission* (FEC) in 2010 and *McCutcheon v. the FEC* in 2014. In both cases, the FEC lost. In *Citizens United,* a corporation which tried to air a movie on presidential candidate Hillary Clinton, the Court ruled that the FEC couldn't restrict a corporation's right to freedom of speech as a person under the U.S. Constitution. All corporations in the U.S. are since allowed to spend any amount of funds to support or oppose any political candidates but not as campaign contributions; however, in *McCutcheon v. the FEC,* corporations are free to spend an unlimited amount of aggregate money on campaign contributions to any number of candidates they want. These two cases amount to a "one-two punch" in favor of corporations to literally purchase whatever candidate they want to be in any political position, including the presidency, and crush any opponents. Corporations are free to spend as much as they need to on campaigns for the politicians who are essentially owned by them and thus work for them to further corporate special interests. "Elections are a commodity for sale and the price is being driven up so that only the very rich [the corporate] can afford them."[13] Democracy has become an obsolete illusion.

MNEs also employ the means to confuse the public. Their pundits create endless debate over any unfavorable science or issue. They create craftily named titles under the guise of environmental organizations – front groups - which continually work to abolish all environmental regulations, including the Endangered Species Act. MNEs also greatly compromise some of the legitimate environmental organizations by funding them and installing corporate board members, insuring those organizations don't address certain issues and to create the illusion of effective environmentalism.[14] The status quo of big business such as cattle production, energy creation, banking, wetlands development, and all sources of natural resource extraction is thus further secured.

Corporations have evolved over the past 500 years to the point of writing their own laws, have politicians and the media under their boot, and have free reign to use the Earth in any way they want. MNEs have changed laws so that protesters or activists who act out in the U.S. for the environment can be classified as terrorists with no rights. In Brazil it's even worse: 1,170 environmental activists have simply been murdered.[15] MNEs can ignore environmental laws – if they're even enforced - because the penalties for any illegal actions are limited by law and amount to minor business expenses that generate huge profits and even become tax write-offs – if they even pay taxes.

The effects of corporate-dominated globalism are numerous. The negative externalities of industry are cumulative: climate change, desertification, radioactivity, deforestation, and the general massive loss of species exponentially worsen. The continents are being homogenized with the spread of invasive species through global trade, which is creating a new Pangaea, but today's version is unprecedented since life on Earth began.[16] MNEs are deciding the fate of biodiversity and what goes extinct, as they destroy habitat, patent seeds and other lifeforms, and shut down biological evolution. "Special interests in general see themselves deserving of what should be preserved."[17] "Corporate globalization is leading us to an evolutionary dead end."[18]

The evolution of chartered corporations to MNEs has ended the battle between Marxism and capitalism; they have become the same. Both lead to the concentration of economic power in unaccountable, centralized institutions; both create systems destructive to those of the living Earth in the name of economic progress; both produce a disabling dependence on mega-institutions that erodes the efficient functioning of markets, governments, society, and spiritual well-being.[19] By means of personhood, longevity, deregulation, and technology MNEs have established a borderless, global economy more powerful than any national government, control the media, steer governments by owning politicians, and form legislation with an overwhelming amount of lobbying. ("In 1998, there were more than 38 registered lobbyists and $2.7 million in lobbying expenditures for each member of the U.S. Congress."[20]) Overall, this has caused an organized, global upper class of people that run the world where nationalities have little relevance, the lower classes are of even less concern, and non-human species are not even considered. It is the new form of dictatorship. By turning away from morality, corporations have sunk into a quagmire of greed, increasing a society that celebrates fraud, theft, and violence.[21]

The corporate world has negotiated and conceded fragmented wilderness areas, parks, and refuges to be preserved that are too small for the genetic evolution of species, and MNEs could have the status of those precious vestiges politically overturned at any time; for now, citizens may enjoy them for newly created fees, while the practices of MNEs otherwise scrape life off of and out of our Earth without obstruction. Corporate-created coalitions fight any issue that might cut into profits, turning the damaging issues around with discrediting "experts" and "news," or cut off funding, contradict science, and question any corrective action - and then dismiss the promoters of public interest as greedy opportunists pursuing narrow special interests.[22] For instance, in a speech thirty-six years ago, representing the corporate attitude that has only grown, Raymond Momboisse attacked environmentalists by saying their "selfish... motivation [and] ability to conceal their true aims and lofty sounding motives of public interest, their indifference to the injury they inflict on the masses of mankind, their ability to manipulate the law and the media, and most of all their power to inflict monumental harm on society"[23] is, rather, exactly what MNEs are guilty of. Even worse, on behalf of Shell Oil, the Nigerian government actually hanged Ogoni tribal leaders, namely Saro-Wiwa, for protesting Shell's devastation of their Nigerian homelands.[24]

"In 1998 when Exxon and Mobil Oil merged they formed the company with the largest assets of any existing industrial corporation."[25] In the first decade of the 21st century, MNEs of the oil oligarchy were making profits of $300 million per day, profits never known in history. "Virtually all of the benefits of America's productivity gains went to... the wealthiest 10 percent who now own almost 90 percent of all business equity... bonds, and... stocks.... [Seventy] percent of world trade is controlled by just 500 corporations... [and] the world's 200 largest industrial corporations, which employ only one-third of 1 percent of the world's population, control 28.3 percent of the world's economic output."[26] As a natural progression, wealth accumulates and becomes political power; such control has been going on in one form or another for thousands of years in the DC. The difference is that control is now planetary in scope, and corporate rule is just the new name for it, rather than monarchy, Rome, or Church.

Since the 1980s, corporate competition has been a greedy race to the bottom. In what is known as corporate cannibalism, predatory schemes have targeted healthy companies, which were seen as weak because their finances were "committed to investing in the future; providing employees with secure, well-paying jobs; paying a fair share of local taxes; paying into a fully funded retirement trust fund; [and] managing environmental resources responsibly."[27] As healthy firms were gutted, MNEs were also moving according to new trade agreements. In 2004, although 210 of the world's 500 largest corporations were headquartered in Western Europe, 199 in the U.S., and 80 in Japan... MNEs of the OCCs moved to Mexico after GATT and then to China (when it joined the WTO immediately after the chaos of 9/11), as wages averaged $21 per hour in the OCCs, $2.36 per hour in Mexico, and $0.47 per hour in China.[28] Corporate executives are exorbitantly rewarded "for contracting out their production to sweatshops that [pay] substandard wages, for clear-cutting primal forests, for introducing labor-saving technologies that displace tens of thousands of employees, for dumping toxic waste, and for shaping political agendas to advance corporate interests."[29] Even more, natural resource extraction has gone into nearly unrestrained high gear, and species loss grows accordingly severe.

Corporations have won; they have performed a coup d'état over every relevant government necessary to do so. Due to the demand from the number of people in the world (which fuels the MNE's enormous profits), boycotts, protests, petitions, letters, initiatives, etc. against MNEs have very little effect. Individual corporations acquire net profits larger than the entire economies of most countries. Politicians work for MNEs from the president on down, or their political careers are over. MNEs are writing thousands of pages of laws in their favor for their politicians to pass. (The agreement that set up the WTO, for instance, was 25,000 pages long). Many individuals work as top politicians to create laws in favor of the MNEs they worked for and then go back to work in top positions for the MNEs they helped as politicians. MNEs control what we eat, grow, and receive as news; they control the computers that count our votes, the energy we use, the vehicles we drive, what we buy, and our health care. A system has been created "that is now beyond the control even of those who created it."[30]

While people have the illusion of freedom, they have no say about war, taxes, foreign policy, health care, energy policies, environmental protection, development, water usage, or biodiversity, etc. We no longer have senators or other politicians that represent some state so much as representing Goldman-Sachs, Shell Oil, Raytheon, Pfizer, or Weyerhauser and hundreds more. "The divide is not between Republican and Democrat. It is a divide between the corporate state and the citizen."[31] Politicians are concerned about representing the corporations that sustain them and no longer need to concern themselves about the people of their districts. "Corporations are not concerned with the common good. They exploit, pollute, impoverish, repress, kill, and lie to make money. They throw poor families out of homes, let the uninsured die, wage useless wars to make profits, poison and pollute the ecosystem, slash social assistance programs, gut public education, trash the global economy, and crush all popular movements that seek justice...."[32] "The corporation is a true Frankenstein's monster – an artificial person run amok, responsible only to its own soulless self."[33]

What goes unsaid is that our Earth absorbs it all, and *species have entered into an era of mass extinction.*

DENIAL

THE DC IS FUNDAMENTALLY based upon denial, but since it has taken on so many forms, it has become such a normal part of our lives that we don't even know we're in it. Forgetting that you're living in denial has become as easy as forgetting you're breathing the air you live in. Dozens of forms of denial exist. Some examples are: diversion, distraction, evasion, negation, and withholding. All denial is various forms of lying. Societal institutions, the media, and large corporations use them all.

Special interests include large corporations and the politicians and know-it-all professional deniers they own (media commentators) and religious organizations. Since special interests are the ones in power who influence legislation and social consciousness to their own advantage, they think the system is working just fine, no change wanted, needed, or even tolerated. To keep it so, they will disparage anyone questioning their system with accusations of dishonesty, incorrectness, bias, sacrilege, pessimism, elitism, emotionalism, alarmism, and misanthropy. They will use several denial mechanisms to spin the truth around so much that they become overt liars. Special interests will ignore evidence they don't want exposed, such as not even reporting on radioactive nuclear accidents, or deliberately interpreting evidence falsely, such as focusing on the four glaciers in Alaska that are growing rather than the 500 that are melting.[1] However, they are so good at bullying debate that the truth will never prevail with them, but a lot of people are willingly convinced by these spin artists because denial makes them comfortable and allows them to not have to think. "Facts no longer exist. Everything is a 'point of view' now."[2] Another denial tactic is the creation of self-serving myths such as the one used by U.S. logging companies that say there are no jobs left because of the environmentalists, not because they've already cut down 96 percent of the forests more than once and now have feller-buncher machines and ship most of the raw logs to Asia. Silence is another denial tactic: just don't talk about it, don't report it, confiscate evidence, and silence the whistleblowers.

Above all, corporate/political denial takes the form of purposeful confusion to cause delay of action; that is their *modus operandi*. Corporations spend billions to purposefully cause even the slightest amount of confusion about the truth to deliberately cause endless debate, saying that not enough evidence is in to prove anything, so that legal process will delay any legal action indefinitely. Therefore, these corporations will never have to change their policies and will thus avoid losing their monopolies of power and profit, allowing them to continue without change for their own benefit at the expense of all other life on Earth. Paying fines and legal fees is much cheaper and even a tax write off. The gigantic Chevron/Texaco oil spills in Ecuador that have been tied up in court for twenty years is a perfect example, as is the NFL's stance on the court battle of professional football causing brain damage[3], or the Catholic Church's incredible, overt lies and denial over the criminal sex-offender priests for decades.[4]

The DC has been manipulating events with dishonesty for thousands of years, all predicated on denial. Honest, trusting people experienced fatal naivety long ago by not realizing how the DC operates. The Trojan War comes to mind, whether factual or mythic, the lying treachery of the "gift" of the wooden horse won a war and established an accepted and even honored way of doing things. How is this any different than honest businesses being victims of hostile takeovers or corporate cannibalism in today's stock markets, or of treaties with Native Americans that were never intended to be honored by the DC before they were even signed? Different forms of denial have been celebrated as successes forever; it's the way entire civilizations have gained unchallenged power. The dishonorable and deceitful have become the winners, and taking all the Earth's resources as much and as fast as possible to stay ahead has been a central ingredient of their success. Forget cooperation, increase your numbers, kill everything and stay alive no matter what it takes... and so we have a mass extinction going on, one that most don't even know or care about. There is massive denial going on about our human-caused ongoing mass extinction, personal denial that it's normal, that nothing can be done, that it's just human nature, and that it's even ok.

Though it has been assumed for centuries that humans are rational beings, a good examination would show that rationality is more often a form of denial used for emotionally-induced selfish gains. Rationality is often used as an unconscious convenience to overshadow our healthy and honest emotional nature that knows right from wrong if we were to admit the truth, but rational denial disallows us from admitting the truth, which allows the DC to do highly unethical things.

The behaviors of denial on a personal level include not being emotionally or mentally present, molding one's self in order to fit into society so as to avoid being marginalized or humiliated, choosing psychological/emotional comfort over being rightfully alarmed (a false sense of security), unwillingness to change, addiction, over-work, choosing ignorance, self-absorption, and procrastination. They are all forms of lying to one's self.

In regards to the HOME, the voices say, "What's that got to do with me? It's not going to happen in *my* lifetime," denying that it already is happening. Or, "Extinctions happen all the time. There's been several before we ever came along. It's no big deal. Life'll go on. It's all good...." Or, "I don't even wanna know. Don't tell me. Don't show me. I know it's bad. Those images (of animal abuse for example) will ruin my meditations...." Delay is another denial that just keeps putting everything off and is a form of hopelessness, powerlessness, self-pity, or overwhelm, and it's much easier than doing something, though its reward is depression or worse.

More and more, there is a growing denial taking place using "spirituality." "Don't focus on the negative; it will only increase it. We don't have to do anything; God will take care of it. Just envision a better world, and to do so, you can't have all those bad images or negativity in your head. It's the Rapture coming; it's all God's plan. Prepare for the End Times or to ascend; doing anything else is pointless and a waste of your lifetime." And so more find ways to allow the HOME to occur.... Maybe they'll get to the Pearly Gates and God will shut them out and ask incredulously, "Why didn't you do anything to help save

My creation!? Why'd you cause and allow the Paradise of Earth to go to ruin? To love is to care. Go back down there and reap the rewards of your own actions and inactions till you get it right. I can't let you through the Gates until you prove you won't ruin this place too!"

If you're bothered by not having a purpose in life, there's a huge cause right in front of your nose to help heal and save Earth's beauty and avoid the worsening of the ongoing mass extinction. Involvement in any form is a plus. End your denial that everything's normal, ok, won't affect you, and fine. Your denial is not even an act of self-love let alone worldly love; even your so-called spiritual practices may be an act of denial. Apathy is the opposite of love, not hate or rage or fear; anything more than indifference is more loving than your obsequious denial.

CLIMATE CHANGE

I N THE PAST 100 years, the average temperature of our Earth's climate has gone up by 1.6⁰F. This is all agreed upon, but what's the big deal? We minimize the fact because we think anything can handle a degree or two, but no one has shown the math.

The average surface temperature of our Earth's climate is 59⁰F, meaning that all the combinations of all the temperatures throughout all the seasons of everywhere on the globe is that average degree. To raise the average by 1.6⁰F, however, an equivalent of nineteen extra days of 90⁰ weather are needed out of the annual 365, and this is actually what has happened over the past century.

Since 1900, summer around the world is coming at least a week earlier per year, and summer is lasting at least a week later per year. Polar ice is breaking up two weeks sooner than it used to. In the northern hemisphere, southern birds have shifted their ranges northward by 1.25 miles per year. In China, the growing season has increased by fifteen days per decade (1.5 days per year). In Hudson Bay, Canada, the annual ice breakup occurs about 2.5 weeks earlier than it did forty years ago.[1] Entire ecosystems, with all their biota and annual life rhythms, are moving an average of 5.5 feet per day toward higher latitudes in both hemispheres to keep up with their accustomed temperatures[2], and forests are migrating up to eight feet higher per year in altitude.[3] Snowmelt in much of western North America is beginning ten to thirty days earlier than previously. Since 1986, wildfires have burned 600 percent more acreage, the duration of the wildfire season has increased by more than two months, and large fires are lasting an average of 500 percent longer from 7.5 days to 37 days.[4]

At 5.5 feet per day, or 28.5 miles in a 75-year human lifespan, a person may live in the same spot for three generations (25 years each) and not notice some of these changes, as most biomes cover much more than 28.5 miles in latitude, so the southern edge of the biome (in the northern hemisphere) may not have passed this person's home in their lifetime. In other words, in the scope of a human lifetime, it makes it easier for some to deny climate change is happening. However, as just one example of hundreds, in Australia in 2009, 173 people died from a heat wave with the worst bush fires ever seen, and 4,000 large fruit bats fell dead out of the trees in city parks. Every major city went on water restrictions.[5]

Accurate global temperatures were first recorded in 1880. Since then, nineteen of the twenty warmest years have happened since 1980; the 1990's became the warmest decade on record; 1998 was the warmest year ever, and then 2005 surpassed it, and then 2010 surpassed that, and then April 2014 tied the record; July 2006 broke 2,300 daily temperature records in the U.S.; record highs for each individual month have occurred in the U.S. since 1997; and thirty-nine consecutive Aprils (since 1976) and 350 consecutive months (since 1985) have had global temperatures above the twentieth-century global average.[6] Over the past sixty years, average temperatures in Siberia and Alaska have risen by the equivalent of an extra twenty-eight days per year of their highest-degree weather, or almost 5^0F. Since the 1800s, humans have added 600 billion metric tons of carbon to the atmosphere and add billions of tons more per year, rising by at least 6 percent annually.[7] "In the second half of the twentieth century, the world's oceans heated by 0.7^0F"[8], an immense amount of absorbed heat. And the thickness of Arctic sea ice has been reduced by up to 40 percent.[9] In general, higher latitudes have warmed more than the tropics (almost double), higher altitudes more than at sea level, nights more than days[10], winters more than summers, and it's warming over oceans more than over land.[11]

In West Antarctica, temperatures have risen almost 11^0F in the past fifty years[12], and the ice sheets of the Pine Island and Thwaites Glaciers (and others) are in irreversible retreat. Consistent, record speeds of glacial movement have also been measured at Greenland's Jakobshavn Glacier, moving three times faster than any glacier ever known: 10.5 miles per year or 151 feet per day. In twenty years, glacial melt in

Greenland has gone up 632 percent to 215 billion tons a year (bty), and West Antarctica's melt is up 490 percent to 147 bty.[13]

All over the Earth biota is moving further north in the Northern Hemisphere and further south in the Southern, or it is moving further up in altitude to attain the cooler norms it has become evolutionarily accustomed to. Similarly, ocean biota is moving deeper and toward the poles for the same reason. Even forests can migrate up to a half a mile average per year. However, all this adaptive movement is not a smooth, linear process; it is more erratic with declines, jumps, stops, and starts. For biota, the candle is burning at both ends since it is warming where they are, but also warming faster in the cooler places they're headed.

Non-domestic species don't have heaters and air conditioners to regulate their environments. That humans do is partly why we have become so out of touch with nature and can more easily deny climate change. Nor do we have to live in the ocean and breathe or live off of a far more acidified environment. The others don't have doctors and medications and air or water filters or indoor environments shut off from the world with doors and windows, and they're in it at all times, 24/7. If you've ever been sea-sick out on the open seas, you know the feeling of having no escape, and it's not going away and it's going to go on forever. Every minute is relentless. Warming today is taking place about twenty-five times faster than it ever has (previously it happened at about 1°C per thousand years). This means biota will have to migrate and adapt that much faster or die, both at the fastest pace ever in paleontological terms. Humans may be unaware of it, but biota is in complete chaos out there, fighting for its very existence, with so much going against it. For instance, as entire biomes migrate, preserved lands will likely become uninhabitable for the animal species they were designed to protect, and there is no room for them outside the boundaries. Today's ecological and related sciences predict that at least one-quarter of Earth's species will be at risk of extinction by 2050 just from climate change. Warmer temperatures also allow insects, disease, and weed species to flourish and overtake habitats.

We are just at the beginning of our carbon debt coming due. Like a coal-fired freight train, the Industrial Age was just idling for a hundred years while its engines got stoked before it could start pulling the world along; the past fifty years it was only just getting its momentum going, just starting to gain speed; even yet, the engine continues to be fueled under more "business as usual" - and this train has no brakes.

The average CO_2 level in our atmosphere has been 280 ppm for eons, as accurately proven by ice-core samples taken from Antarctica, Greenland, and Alaska, some with actual air bubbles trapped in the ice for hundreds of thousands of years. Currently, the atmospheric CO_2 level is at 404 ppm, a 44 percent increase since 1850, and we are adding at least 2 ppm per year. Scientists have proven that atmospheric carbon has not been this high in 800,000 years, which is as far back as they can physically measure, and speculate that it has not been this high for up to 55 million years. "Three or four degrees hotter is generally equated to an atmospheric concentration of 450 parts per million of carbon dioxide." – Gwynne Dyer.[14] At the rate we're going, by 2050 there will be 500 ppm of carbon in the atmosphere. It's figured "that an 85 percent cutback in CO_2 emissions would be required to stabilize atmospheric CO_2 at 450 ppm. ...Stabilization at the 1990 level (350 ppm) is virtually impossible, because it would require negative net CO_2 emissions sometime in the future."[15] First, however, people must make leaders realize and accept that policies continuing business as usual must be truly revolutionized.

Twenty-five percent of the carbon released into the atmosphere stays there for 500 years, but carbon is only half the problem; methane contributes to most the other half. Methane on Earth is and was largely created by bacteria, and it is fast becoming a much greater concern than CO_2. In fact, CO_2 may simply be the initiatory factor for the release of methane by warming the climate just enough to trigger methane's release. Methane breaks down in the atmosphere much faster than CO_2, but over its first twenty-year time period, methane is sixty-three times more potent a greenhouse gas than CO_2, and it has increased its presence in the atmosphere by up to 450 percent since pre-industrial times. There is suddenly more methane in our atmosphere than in a minimum of the past 600,000 years. As the Arctic sea ice has been receding, the Arctic Ocean has been warming along with the surrounding land masses, releasing methane from the thawing tundra and sea beds; in fact, great plumes of methane are being released from the ocean 100 to 1,000 times larger than ever observed, and parts of the Arctic Ocean are bubbling with methane like seltzer water. The thawing tundra also covers over twenty percent of the Earth's land mass.

There are four other new sources of methane release. Natural gas is mostly methane. Fracking (hydraulic fracturing) is the creation of natural gas wells and the use of chemically treated water to fracture underground rock that is permeated with natural gas. Five percent of thousands of these new wells leak methane from the onset, and experts say that eventually 100 percent of the wells will leak.[16] These wells are being drilled all over the world.

Indonesia, home of the endangered orangutans, has now lost about 99 percent of its original tropical forest and has been draining its peatlands, largely for the planting of palm oil plantations. In the process, methane is being released from the large, drying, ancient deposits of peat, releasing enough to make Indonesia the third-largest contributor of greenhouse gases in the world after the USA and China.

Third is the release of methane by domestic ruminants: livestock. This is considered new because, essentially without predators, their numbers have vastly increased over any other time in history in order to supply the exponential growth in human population and wealth, greatly throwing off the balance of nature. Cows, sheep, goats, and any other cellulose-consuming animals are belching enough methane to actually make a difference. Along with the gaseous releases of over seven billion people and their pets, it adds up to a surprising amount of greenhouse gas emissions – some say 23 percent when you factor in the production, transportation, and packaging it takes to feed them all.

Incredibly, termites are a fourth mentionable component of methane contribution. As the tropical forests have been obliterated, termites have proliferated in unbelievably exponential numbers with gigantic nests, all excreting methane as they devour the planetary wreckage of these forests. "In Brazil [for example], deforestation accounts for as much as 70 percent of the country's total greenhouse gas emissions...."[17], including the burning of trees.

When all told, all of these extra greenhouse gas contributions to our atmosphere are being produced only by mankind: the burning of coal for electricity and heating, the use of oil and gas for transportation, the clearing and burning of forests and even the termites that follow, the drainage of peatlands, the thawing of the Arctic zone with its resultant release of methane from the permafrost and sea beds, the overabundance of belching domestic ruminants, and the prevalence of fracking for natural gas.

There is no debate in the scientific community about if climatic warming is happening; endless debate exists only for those who choose to believe the professional deniers that represent the special interests that make and save billions of dollars by maintaining the status quo by keeping the public muddled in confusion and argument. The evidence is now overwhelming that Earth's climate is warming due to the release of greenhouse gases from human activity.

Slowing the momentum of climate change down is like pushing on an extremely heavy weight, a huge hundred-ton lead orb. It took a lot of energy to get that thing moving, but now that it has started, it'll take at least as much energy to stop or reverse it. That is one aspect of the environmental debt, the greenhouse gas debt. We have already permanently altered the Earth, and there is much more to come.

ENERGY

THE AMOUNT OF ENERGY humans use is infinitesimal compared to how much comes to us from the Sun. Any direct light our Earth receives from the Sun contains 1.35 kilowatts of energy per *square meter*, enough from one hour of world receptivity to supply more than the entire human population can use in a year. A total area of only 3600 square miles (60 by 60 miles) of solar energy could turn enough steam generators to supply all the electricity demands of the U.S.[1] One type of solar technology called concentrated solar thermal power (CSP) focuses the Sun with mirrors to pressurize fluids that turn engines or turbines to generate electricity the same as any other source but without carbon or radioactive emissions.

The amount of energy humans use is also miniscule compared to how much is available from our Earth's magnetosphere. As part of that system, lightning occurs an average of 100 times per second somewhere on Earth; in one year, over 22 million lightning bolts are cloud-to-ground strikes. In this way, natural law maintains the exchange of electrical balance between the partially-molten iron core of Earth, a spinning dynamo, and the magnetosphere generated by it.

Nikola Tesla, one of the greatest geniuses of all time, understood this and invented how to use the Earth's naturally occurring electromagnetic energy. He invented the Tesla Rod, a device that was planted in the ground and wirelessly collected and transmitted electricity directly from the magnetosphere. Tesla also invented AC electricity, which we've used for over 100 years, and the Tesla Coil, which could amplify our use of AC electricity by at least 80 percent. He proved there was no need to burn fossil fuels for transportation or for heating or for electricity – that the magnetosphere could do it all - but his benefactors had other plans.

Born in Yugoslavia in 1856, Tesla lived his entire adult life in America, passing in 1943. Nikola patented over 1,400 inventions, beginning at a time when the world only had horse-drawn carriages or steam-engine trains and ships for transportation; only had whale oil, wax, or kerosene flames for lighting; and had no electricity or indoor plumbing. It was a time when the structures for power and energy were being formed for the budding new world. America was basically settled and ready to jump into the new age that technological inventions were providing. The traumatization of the Civil War was ebbing after a generation's time, the Indian Wars were just over, and the DC had the best opportunity ever in human history to grow in fertile ground in the best of conditions.

After the notorious Thomas Edison, founder of the General Electric Company (GE), had hired Tesla, only to rip him off later by dishonoring his agreement to pay him $50,000 in wages (over $1 million in current value), George Westinghouse and J.P. Morgan eventually became Tesla's two main financial supporters. George Westinghouse, who had befriended Nikola, bought the Tesla Rod patent directly from him, resulting in the suppression of this technology to this day. Expanding on his Tesla Rod technology, Tesla had a huge tower built at Wardenclyffe on Long Island, New York, to develop sending wireless electricity to any part of the world. J.P. Morgan soon withdrew his funding for all of Tesla's research, Tesla's main lab was burned to the ground with most of his records and research, and his Wardenclyffe Tower was destroyed.

In the government-funded public schools, students never learn a single thing about Tesla but hear plenty of glorifications about Edison for having invented the light bulb. Tesla invented wireless light bulbs, without batteries, that lit up in his hands or when he planted them in the ground. He invented x-ray technology, patented computer algorithms that were used decades later when computers were invented, and the "star wars" defense system planned in the 1980s was based on Tesla technology. When he died, the American government confiscated all his research.

The destruction and suppression of Tesla's technology is one of the most monumental tragedies and cover-ups of all time. Westinghouse went on to create the electric power grid still in use today, J.P. Morgan supplied all the copper for that grid at the time, and John D. Rockefeller supplied the oil and gasoline for the world's new transportation and industrial infrastructure. It didn't have to go that way, but the "robber barons" seized control and chose to establish the energy technology and systems we still use today. Chase Bank, one of the largest international banks in the world, is a direct descendant of J.P. Morgan. The monopoly of Rockefeller's Standard Oil Company got broken up by anti-trust laws into Exxon, Mobil, Texaco, Chevron, Atlantic Richfield (ARCO), and several others only to all merge again decades later into today's familiar oil MNEs such as Exxon/Mobil, Chevron/Texaco, BP/Amoco/ARCO, and others.[2] The robber barons created their own monopolies and became the richest men in the world, the equivalent of today's multi-billionaires, all under the guise of providing for the benefit of humanity; however, if that were truly the case, they would've let Tesla's technologies flourish, but there would've been no money, no power, and no control in it for them.

Had Tesla's energy paradigm been implemented, we would not be using coal-fired or natural-gas-fired energy plants, we would not be using fossil fuels to drive cars, and we would not have built nuclear power plants just to create steam to generate electricity. There would never have been the giant oil spills we continue to have; the world would be free of so much radioactive waste and leakage from Hanford, Chernobyl, Fukushima, and many others we've never even heard about; the climate would not be heating up from human-caused atmospheric carbon dioxide; the acidification of the oceans would not be occurring; the Polar ice would not be melting; we would not be fracking for natural gas; there would be no wars over oil; we would not be erasing and supplanting the tropical rainforests for sugarcane and palm oil for biofuel; corn for ethanol would not be decimating the Great Plains of America, its rivers, and the Gulf of Mexico; and millions of innocent, precious species would not be threatened with the eternal erasure of extinction.

With over $200 trillion worth of remaining oil reserves, the ones in control of these are not going to voluntarily give it up and are doing everything they can to keep the new technologies from widespread, public discovery. Over 5,000 "free energy" patents have been suppressed by the U.S. government (which is controlled by MNEs). Adam Trombly built a dynamo that accessed electricity right out of the air, enough energy to transform the entire Earth; the first Bush president undermined this technology and had the generator confiscated by a government raid. John Bedini had his Tesla-like generator suppressed; he was attacked in his lab and told to quit. Canadian John Hutchison used Tesla's theories to counter gravity and make objects float; he was raided and the equipment taken by the government in 1978, 1989, and 2000. Dr. Eugene Mallove, an engineer from MIT and Harvard, was beaten to death in 2004 for his "free energy" inventions. The laboratories of hundreds of other new-energy inventors have been treated the same: Dr. Brian O'Leary (former NASA astronaut and Princeton Physics Professor), Paramahamsa Tewari, Shiuji Inomata, and Floyd "Sparky" Sweet....[3]

Another huge scam is the production of ethanol, used for supplementing gasoline. People were made to believe it would be better for the environment and would help solve "the energy crisis," but it has done neither. This is beside the point that we already have the technology that would make gasoline-powered vehicles obsolete: very successful, GM-EV1, fully-electric cars were taken out of production in January, 2000, and all destroyed by 2005[4]; diesel engines could run on 100 percent biodiesel fuel; and suppressed Tesla energy patents could be brought to light to replace even these. Even gasoline cars could easily get over 100 miles per gallon as many tests have proven for decades.

Henry Ford designed his first vehicles to run on pure ethanol. This is because the country was without an infrastructure of roads and gas stations, and it would enable farmers to distill and provide their own fuel. Mr. Ford also made electric cars, which were the only ones his wife and daughter drove. Over half the cars in 1920 were electric cars. The final dominance of using fossil fuels for transportation took place during Prohibition in the 1920s, which went as far as even adding an amendment to the U.S. Constitution prohibiting the production of any alcohol, including ethanol, with a lobbying effort largely funded by J.D. Rockefeller, America's first billionaire. Nine years later, once the gasoline infrastructure was in place, and only after the idea of electric and ethanol cars was defeated, the Prohibition amendment was repealed.

In 1973, the American government enacted legislation to begin ethanol production to assist in reducing toxic lead emissions from gasoline. Five years later, it began subsidizing corn ethanol production. Twenty-nine years later, in 2007, the disastrous Energy Independence and Security Act was passed, which called for 36 billion gallons of biofuels to be produced per year by 2022, a 400 percent increase over 2008's nine-billion-gallon production. The corn-ethanol industry then went into high gear, and ethanol became 99 percent of all biofuel production in the U.S.[5] Biodiesel, a much healthier option, which could *easily* replace the entire heating-oil supply, was excluded.

The U.S. is the world's largest producer of ethanol, mostly from corn, but the U.S. ethanol industry only exists in a highly-manipulated market: if Congress did not provide it with subsidies, tariffs, and tax credits, the ethanol industry would cease to exist in an open market. Consumers and tax payers are paying far more for ethanol than is realized. U.S. government handouts to the corn-ethanol industry amounted to almost $12 billion in 2007, with subsidies of $7 billion and tax credits of $4.7 billion.[6] To further keep the ethanol industry propped up, a tariff of 54 cents per gallon was imposed on ethanol imports to prevent cheaper ethanol from entering the U.S. from Brazil.[7] (Brazil is the world's second largest producer of ethanol, from sugarcane, which has greatly contributed to destroying the Amazon rainforest.) However, this $12 billion amount of corporate welfare in 2007 does not reveal the even higher hidden costs of producing ethanol.

Corn is at the bottom of the list in the ability to create energy; corn trails soybeans, the next lowest rival, by three times, and palm, the most efficient biofuel food crop, by thirty-five times.[8] Corn is the least efficient of all crops in producing ethanol, but it is great for the short-term economy: corn requires more pesticide, herbicide, and fertilizer than any other food crop grown.[9] Furthermore, corn also requires more water than any other crop (except rice), causes the highest amount of soil erosion, results in less crop rotation, and chemicals all the way from farms in Iowa and Minnesota flow down the Mississippi River into the Gulf of Mexico, increasing the Dead Zone - the part of the Gulf where nothing can live - by up to 18 percent per year, killing off large amounts of marine food resources.[10]

Because ethanol production increased the price of corn, farmland devoted to raising corn increased in 2007 by over twelve million acres, to 92.4 million.[11] By 2008, the corn crop equaled covering every square foot of the state of California in nothing but corn - no cities, lakes, roads, forests, National Parks, nothing. Since one-third of U.S. corn currently goes to producing ethanol, meeting the 2022 goal set by law requires quadrupling the amount of land allocated for the exclusive use of growing corn for ethanol; this would take up every square foot of an area larger than the states of Iowa, Illinois, Indiana, and Ohio combined or that of California plus over 20 percent just for ethanol. Plans call for 200 percent more corn to be devoted to ethanol than corn to feed humans and livestock; additionally, the U.S. is the largest provider of food corn to the rest of the world, exporting 15 percent of its annual yield. Much of the corn, of course, will use the Midwest's rapidly diminishing Oglalla underground aquifer, which is irreplaceable: once gone, *no more water for the Great Plains, "the breadbasket of the world."*

The proponents of ethanol claim that as much as possible should be produced because not only will it supposedly help "wean us off foreign oil," but compared to gasoline, ethanol has a higher octane rating (112 to 87) and a higher oxygen content, making it a cleaner burning fuel.[12] While both of these facts presume that ethanol is healthier for the environment, actually the exact opposite is true.

The costs of corn ethanol production are far greater than the benefits. To create thirty gallons of ethanol – enough to drive about 600 miles - 660 pounds of corn are needed, and over 1,000 gallons of water are needed to create just one gallon of ethanol.[13] Though ethanol was supposed to decrease the need to import oil, it takes 46 percent more oil to produce it, actually increasing oil importation for the U.S.[14] Ethanol also needs to be shipped by truck, barge, or train, using more petroleum, since delivery through pipelines causes too many impurities. Additionally, though ethanol burns cleaner than gasoline, it emits other noxious gases that gasoline does not, such as peroxyacetyl nitrate (PAN), acetaldehyde, and alkylates.[15] For all this, various sources argue as to whether ethanol even increases gas mileage or not.

In 2011, 134 billion gallons of gasoline were used in America and 12.95 billion gallons of corn ethanol were consumed, which was only 9.66 percent of total U.S. oil consumption, yet it took 33 percent of the U.S. corn crop[16] and *13 **trillion** gallons of water*. Annually, *36 trillion* *gallons of pure water* will be used to produce the goal of 36 billion gallons of ethanol. As other crops are sacrificed to make room for corn, a global price increase is caused for corn's substitutes such as wheat, barley, and others, which also raises the price for livestock feed. The UN food expert, Mr. Jean Ziegler, in his 2007 speech to the world, correctly called the production of food for fuel (ethanol) "a crime against humanity."[17] He wasn't even considering the water use, chemical pollution, habitat destruction, and loss of species.

Biodiesel is another form of biofuel but is very different from ethanol. Ethanol is a distilled alcohol, whereas biodiesel is made from vegetable oil or animal fats, and biodiesel can fully replace gasoline, diesel, or ethanol.

Named after its inventor, Dr. Rudolph Diesel, the first diesel engine was demonstrated in 1900 at the World Exposition in Paris, France, two years before Tesla's AC electricity was introduced at the Chicago World's Fair. The first diesel engine ran on 100 percent peanut oil, and this original engine still runs today on pure biodiesel.

Diesel engines are the most dominant engines in the world and power trains, semi-trucks, ocean liners, power plants, and comprise fifty percent of new European car sales.[18] Biodiesel can also be used for domestic heating oil. There are many reasons why biodiesel is not widely available, not the least of which are from influences of the international oil oligarchy that does not want to have biodiesel fuel replace any part or all of its crude oil market.[19]

Research in the science of biodiesel has discovered certain types of algae that are the most abundant source of raw material ever realized for this fuel.[20] Policies of President Carter began major efforts of this research in the late 1970s, but this research was shut down as soon as Reagan entered office. Biodiesel research has long since lost all of its governmental funding while all investment instead went towards ethanol development. However, the promise of making biodiesel fuel from algae is astounding. Biodiesel algae production can use land that is not farmable, use algae species that are not edible, and can produce thirty times more biodiesel than the best oil seed crops[21] such as palm, whose plantations are rapidly replacing and destroying the tropical rain forests that are home to the endangered orangutans, threatening them with actual extinction. Algae farms can be linked directly to the CO_2 emissions given off by power plants and to non-potable waste water from treatment plants, which both make algae thrive exponentially.[22] Biodiesel fuel is biodegradable, reduces net CO_2 emissions by 78 percent compared to diesel and 650 percent better than ethanol[23], emits no sulphur, and returns a net 4.5 units of energy for every one unit of fossil energy it takes to produce it.[24] The benefits of biodiesel are numerous, including being a sink for CO_2 when biodiesel crops regrow, as they draw as much carbon out of the atmosphere as they give in energy.

Without government funding, advancements in biodiesel have come slowly from the private sector. The first generation of biodiesel comes from food crops (as does ethanol) consumed by humans. The second generation extracts biodiesel from crops not consumed by humans, such as camelina, a weed shrub that grows in North America, and jatropha, a weed tree that grows in Southeast Asia and Africa. Both plants are being processed and sold as biodiesel jet fuel to commercial airlines such as British Airways, Qantas, Continental, and Northwest Airlines.[25] A third generation of biodiesel has entirely moved away from using any farmland or crops that could produce food for humans or livestock, such as neochloris oleabundans, an algae comprised of 56 percent oil that produces 800 times more oil than the same amount of corn.[26]

Pyrolysis is another energy technology that could add to the replacement of fossil fuel's dominance. Pyrolysis is a sustainable process of turning our Earth's unending organic matter (waste included) into charcoal through a smoldering technique, extracting the moisture and particulate from the unburned smoke, cooling the smoke, and using the gas that remains as fuel. The charcoal can also be used as fertilizer to replenish farm soils. Pyrolysis was a very big issue in WWII in regards to who was going to control what kind of energy.

Current sources of electrical generation come from coal, natural gas, oil, and nuclear fission. Eighty percent of China's electricity comes from coal, and in 1989, it was building an average of two new coal plants per week. By 2030, "the consumption [of coal] in the United States is projected to rise nearly 40% over 2010 levels. In China, which already burns twice as much coal as the United States, consumption is projected to nearly *double*.[27] (In November, 2014, the U.S. and China made a non-binding, meaningless agreement not to build anymore coal plants after 2030; it was merely a political publicity stunt.) Today, about 95 percent of humans' use of energy comes from fossil fuels, and we are indulging in it at a pace of one hundred times more than can be replenished by our Earth. In the twentieth century, America increased its use of coal by over 300 percent and its oil usage by almost 17,000 percent. Currently, the world uses 89 million barrels (almost 4 billion gallons) of oil per day, which equals the amount of water that flows over Niagara Falls in two hours.

Commercial crude oil was first extracted from the Earth in the 1850s in Pennsylvania. Eighty years later, it was discovered in the Middle East, the Mother Lode of stimulant for the DC.

With the discovery and use of oil, man's industrial or physical work capability was increased by an unbelievable 600 percent; it was the beginning of the HEE, and America led the way. In the 1930s, the HEE got an extra jolt of energy from the Mother Lode, and after WWII control of that energy was established. America took off with it, and the new technologies, especially in agriculture, increased the human population by 350 percent after the war and the introduction of the Green Revolution that fed the world. The establishment of oil as the world's energy paradigm has enabled all the events that have led up to our Human-caused Ongoing Mass Extinction.

Extracting oil from tar sands is a massive strip-mining job entailing gigantic machines that gouge the sands out of the Earth, transport, crush, hydrolyze, and re-transport the new mixture to a chemical extraction plant that removes sand, clay, bitumen, and sulphur, then adds natural gas to get one barrel of synthetic crude oil from two tons of tar sand. There is a huge amount of waste involved: chemical wastes are sent to tailing ponds, sulfur blocks are simply stacked up, and an average of three cubic meters of water and 170 cubic meters of natural gas are used to produce a single cubic meter of tar sands oil. Tar sands oil extraction is an extremely toxic process and has increased Canada's carbon dioxide emissions by over 27 percent, completely ignoring the country's signing of the international Kyoto Treaty to decrease its CO_2 production by 9 percent.[28] The area of Alberta's tar sands mining blight is already larger than the entire country of England and growing.

After over a century of status quo, we think there can be no other way, or that Tesla's technology wasn't real, or that our energy paradigm is so entrenched now that it can never be changed. It has all been a gigantic, 110-year, fraudulent blunder, a blunder of greed, and we need to realize it, admit it, and correct it. The grip of the robber–baron legacy must end or the world dies the ten-million-year death of our anthropogenic ongoing mass extinction. Our world desperately needs a fundamental shift into a new energy paradigm, a paradigm that could have been chosen over a century ago. Whatever else there is to protest about or wake up to, this one issue is the most vital and transcends all others in its potential to avoid the worst-case scenarios of our HOME.

The American hyper-version of the DC led the world into this energy paradigm, and it may be the only one to lead the world out of it. The American government must stop subsidizing ethanol production, stop natural gas fracking, stop its demand for tar sands mining, and definitely stop nuclear energy production and close all those plants down. All of these are extremely toxic to the environment and very inefficient to produce. The American government must invest in true solar energy solutions, including CSP, all the way to amending all building codes to mandatorily install solar panels on every home and office, including retroactively; free the genre of Tesla energy technology patents; invest in biodiesel fuel production; invest in pyrolysis gas production; and mass produce fully-electric vehicles. There is far more economic growth in these fields than in the status quo, *and the status quo is massively killing life on Earth.*

The rule always should have been, *if it's not good for the earth, don't do it*, but this has never been the DC's mindset; we are now beginning the age of obviously reaping the negative results. It's never too late to revolutionize our energy paradigm – and we need to, *now*. Otherwise, we'll continue to create manmade radioactive waste that we still can't even dispense of safely; foul the oceans with crude oil and radiation and acidify it with CO_2; pollute even the purest of waters leftover from the Ice Age in our underground aquifers with manmade fracking chemicals; contaminate the air we breathe, changing its very chemistry; continue to weaken the ozone layer; and much more.

RADIOACTIVITY

THE FIRST NUCLEAR POWER plant in the world was built at Hanford, Washington, along the Columbia River. It supplied the enriched uranium for the bomb dropped on Hiroshima and the plutonium for the bomb dropped on Nagasaki that ended the war with Japan in 1945. By 1985, 103 more nuclear power plants were built in the U.S., designed to operate up to 2025. There are now 435 nuclear reactors on Earth; forty-eight are in Japan, a country the size of Montana; and seventy-two new ones are currently being built, 40 percent (twenty-nine) of those in China.

Humans entered an extremely reckless power struggle in the twentieth century, culminating in nuclear weaponry just seventy years ago. In just one human life span, we have given our Earth a radioactive legacy that will change the rest of Earth's existence forever, and we've already accepted this as normal, acting as if there's *still* no other choice. Maybe seeing the growing amount of freakish human (and animal) birth defects already taking place from changing our Earth's gene pool with radioactive poisoning should be mandatory. The world's nuclear energy plants should be shut down and phased out immediately.

Yes, the discovery and use of nuclear fission ended a period of severe darkness in a war that killed millions, but to this day no one knows how to safely store the waste that nuclear fission produces, such as plutonium, which is radioactive for 500,000 years, or depleted uranium (DU), which lasts for 4.5 *billion* years. The insane mentality that thought nuclear energy was a good idea has burdened the rest of life on our planet with the most toxic poisons known, and they're all manmade artificial atoms. Making electricity with nuclear energy should never have even been an option, but it created an entire industry for the economy and made a few greedy manipulators very wealthy.

Nuclear fission is *not* a clean energy in any sense of the word[1], contrary to what the brainwashed of the DC say, including environmentalists. Nuclear fission "creates more than 200 new, man-made radioactive elements,"[2] the most deadly of them being plutonium. One-millionth of a gram of plutonium inhaled into the lungs is enough to kill a person[3], and there is now over 250 metric *tons* of weapons-grade plutonium in the world. With 1 million grams in a metric ton, humans have made 250 trillion lethal doses of plutonium, enough to kill every person on our planet 350 times. That's conservatively calculating just one type of these artificial, man-made, radioactive elements.

At least 500 tons of depleted uranium were used in the two recent U.S. invasions of Iraq and literally turned into long-lasting dust; DU rocket shells enter a tank through its armor like a bullet through wet clay, exploding with pyrophoric dispersion into the vast, uncontained environment.

Tritium, which readily bonds with water, is radioactive for 248 years; because of a massive leak of 50 trillion curies into Lake Ontario in April, 1996, the lake will be radioactive into the twenty-third century.[4] Cesium 137 is radioactive for 600 years; 400 reactors operating for twenty-five years at 99.9 percent perfect containment still leak enough cesium to equal sixteen Chernobyl accidents.[5] And all lasting radioactive substances bio-accumulate up the food chain in all organisms. As it is, "government regulations allow nuclear plants to routinely emit hundreds of thousands of curies of radioactive gases... into the environment every year."[6] As a murder weapon, only a few hundred nanograms of polonium 210 poisoned a Russian dissident to death (Litvinenko) in 2007; polonium emits radioactive helium atoms into every cell of the body, and it takes just two atoms to disintegrate a cell.[7]

The creation of nuclear fission may already have caused the greatest mass extinction in the history of our world; we may be living only in the grace period of the "radioactive debt" until that debt catches up. With as much nuclear fuel, weapons-grade nuclear bomb material, nuclear reactor leaks, and radioactive waste - about 500,000 metric tons that we don't even know what to do with - it's commonsense to see we have created a very precarious future even if we shut down all nuclear energy operations *now*, and we must.

A wave of radioactive contamination had already spread throughout the world by the early 1960s from nuclear bomb testing. "It is thought that 80% of cancers... are caused by environmental factors...."[8] In the 1960s it was said that one out of five people would get cancer. Now, just fifty years later, it's said that one out of two will get cancer. How long before it's one out of five that *don't* get cancer? Twenty years? Since most cancer deaths from nuclear radiation happen years after the initial exposure, they are never officially attributed to the nuclear industry; the industry is never held accountable, allowing it to continue operations. "It takes up to twenty generations for recessive mutations to... [manifest] as a specific disease... such as cystic fibrosis," and "a total of 16,604 genetically inherited diseases [are] now described in the literature" of the National Institute of Health.[9] This number is now guaranteed to expand. One thing is certain: producing electricity from nuclear fission is the ultimate manifestation of insanity. Some thought it was a good idea because it would generate billions of dollars, and the byproducts could protect us from our enemies. Or is it otherwise, that the real byproduct is the generation of electricity to sell the idea of creating such horrendous weapons....

Nuclear fission has created an entire industry of jobs for those who don't care if they're killing the Earth. It's a very dirty fuel because of what it takes to produce the final product - as if the final product isn't bad enough. To fuel one nuclear reactor for one year, 162 tons of natural uranium must be mined. As it's often in concentrations of four grams per ton, mining 40 million tons of rock is required; this alone uses over thirty times the amount of fossil fuel energy than is generated by the nuclear reactor. The mined rock is then transported, crushed, milled to a powder, and treated with highly toxic chemicals, including sulphuric acid, to make "yellow cake."[10] To enrich and convert it into uranium hexafluoride gas, separating uranium 238 from uranium 235, huge amounts of chlorofluorocarbons (CFC) are used. CFCs are 15,000 times more potent a greenhouse gas than CO_2 and frighteningly effective destroyers of our diminishing protective ozone layer. The uranium 235 gas is made into solid fuel pellets the size of cigarette filters, which fill twelve-foot long zirconium fuel rods. "A typical 1,000 megawatt reactor contains 50,000 fuel rods – about 100 tons of uranium."[11] A reactor with 50,000 fuel rods "manufactures 500 pounds of plutonium a year,"[12] enough for fifty atomic bombs. Exorbitant amounts of fossil fuels are needed to produce all these operations. The massive reactor buildings also need to be built and radioactive waste transported and stored, etc.[13], adding immensely to our manmade causes of climate change.

All this is done simply to boil water to make steam, steam that turns electrical generators the same as coal, natural gas, and rivers do. Furthermore, "nuclear power has been and still is dependent upon [billions of dollars of] government subsidies..." every dollar of which steals from developing better sources of power.[14]

In an agreement that stands to this day, in 1959 the International Atomic Energy Agency (IAEA), which outranks the UN's World Health Organization (WHO), forced an agreement with them that barred any reporting, monitoring, warning, or researching the health consequences of military or civilian nuclear fission use. No studies have been done about Chernobyl, as "the IAEA blocked the proceedings [because] the truth would have been a disaster for the nuclear industry."[15] The same policy is going on with regard to the Fukushima event of March, 2011, which has been spewing radiation into the air, the ocean, and the Earth unchecked since the day *three* reactors (rather than Chernobyl's one) melted down. In June, 2015, authorities finally admitted they do not know where the three 500-pound molten-blob cores of plutonium are, except that they are not in the containment vessels.[16] Fukushima is thus sitting on the equivalent of 150 uncontrolled atomic bombs, the raw presence of the unprotected plutonium unknown. Ruling with an iron fist, the IAEA is in charge of very unscrupulous and institutionalized, official denial.

Most of the reactors in the U.S. and Japan were built or designed by General Electric and Westinghouse. Both companies retain the legacy of the robber barons who buried Tesla's technology so they could profit from controlling the world's energy grid they set up 114 years ago, further entrenching the paradigm of "the way things are."

Chernobyl happened on April 26, 1986. Over 8 million people within 100,000 square miles were dosed with acute radiation, which will remain there for thousands of years. Radioactive fallout occurred throughout Europe where many wild animals and plants are still too radioactive to eat.[17] The ruined Chernobyl reactor is as radioactive as ever, enclosed within a giant, decaying containment building that was designed only to last until 2017. The cost of containing the reactor was the final straw that caused the economy of the former United Soviet Socialist Republic (USSR) to collapse, forcing the end of its governance over the many countries it once controlled.

There have been dozens of nuclear bombs blown into the atmosphere, dozens of major nuclear reactor accidents that are documented, and many more unreported such as radioactive wastes disposed of in very illegal and unethical ways by organized crime.[18] The entire industry is corrupt to the core, from its ideology on down; highly immoral, its manmade poison invisibly infiltrates the very gene pool of our world with the most deadly toxins known in all of Creation.

TIGERS

THE GENUS *PANTHERA* HOLDS all the species of cats that are able to roar: the gorgeous and majestic leopards, jaguars, tigers, and lions. They evolved from a common ancestor similar to a modern leopard that existed circa 5 million BP.[1]

While all species of *panthera* are suffering great losses, most attention is focused on tigers. The tigers' primary habitat spreads across most of Asia, and in the 1970s a shooting spree caused three species of tigers to go extinct, as the last Caspian tiger was shot in Turkey; the Java and Bali tigers were hunted to oblivion; and the Chinese government killed all of the South Chinese tigers in the wild.[2] Now there are only four species of tigers that roam free.

Tigers, perhaps the most magnificent of all creatures to frequent the Earth, are down to fewer than 2,500 adults of all tiger species in the wild.[3] Half are in India, which had over 100,000 a century ago.[4] "Wild tigers have become so rare that all subspecies are endangered, some critically."[5] This includes the Sumatran tiger, the last surviving species in Indonesia, and the Amur tiger in Far East Russia. The legendary Amur tiger, which grows to 850 pounds and thirteen feet in length, has been hunted to less than 500 in the wild.[6] An excellent true story is written about one of them by author John Vaillant.[7]

Every day the existence of tigers comes closer to being solely dependent upon precarious captive breeding programs in zoos, including the famous Bengal tiger; it is there where tigers may go extinct after they suffer dispirited lives full of abuse in artificial environments. If wild tigers are exterminated due to illegal hunting and habitat loss, and all that's left are captive tigers – if they survive - their genetics will be forever altered.

Several other wild cats are endangered besides tigers. Since 2000, a pro-development ploy in the U.S. is trying to have thirty-two cougar subspecies get reclassified into just six. With this manipulation the Florida panther would not require being listed as critically endangered, allowing development to continue in the panther's shrinking habitat.[8] This would mean yet another species would go extinct because of big business maneuvers. The critically endangered ocelot numbers less than 100 in the wild. Its range includes shrinking habitat in parts of Mexico and the U.S., further complicated by the barrier of large border fences. Jaguars all through Mexico and Central America are critically endangered, as they have almost nowhere to live due to the density of the human population. In Southern California and parts of the western U.S., all wild mountain lions have been captured at some point and are now monitored with radio-frequency devices, now quasi-wild. There are only thirty-five Amur leopards left in the wild.

It is the same throughout the world with endless examples, as the countless activities of exponential human encroachment upon the un-tame require ever-demanding protection of humans from such exquisite creatures. These masterpiece species are essential to a healthy environment as apex predators, and their loss is indicative of the harm done to such environments. Their demise is a severe tragedy whose import is as yet unrealized by humankind.

HABITAT DESTRUCTION

H ABITAT DESTRUCTION IS THE number one reason species are going extinct; worldwide, "about 90 percent of the known extinctions of species have occurred because of habitat loss,"[1] and it's humans that are doing it. We have cut down 75 percent of the world's original forests, turned all lands we could into farms, built dams for irrigation and power, scoured the bottom of the ocean to ruination with trawling, poisoned uncounted numbers of rivers and other fresh water ecosystems with chemicals and livestock microorganisms, created over 150 oceanic dead zones around the world with agricultural chemicals, grazed millions of acres of land into deserts with our domesticated ungulates, blown up mountains for coal and other mining, killed millions of acres of soil with chemicals and burning, caused giant radioactive leaks and oil spills on land and sea, blew up atolls with nuclear bombs, created genetically modified species that infected healthy species forever, changed ecosystems by killing apex predators around the world, chemically caused holes in our protective ozone layer, turned radioactive depleted uranium into dust over the Middle East in war to be blown around the world forever, caused the takeover of weed species that invaded after our environmental disruptions, tore the land to shreds with our recreational vehicles where they weren't even allowed, illegally disposed of lethal toxic waste wherever we could, spread fracking pollutants on land and in freshwater aquifers, created numerous toxic superfund waste sites with computer byproducts, covered about five percent of land mass with concrete, created enormous landfills or just dumped our wastes in the ocean, have caused an increasing probability of a mega-release of methane to be freed into the troposphere, and so much more. How many wild animals have been killed in all our wars? Now some countries even illegally plunder animals for their body parts to fund their desperate little wars. And all of this not only continues but escalates.

Of all the things we're doing to destroy our habitat, deforestation and agriculture (including cattle grazing) are the top, all done basically just to house and feed ourselves with the least amount of effort, with the lowest common denominator most conducive to economies of scale. We think we're doing it with the most efficiency: just clearcut the forest and burn the slash, just clear the land and plant crops from hedgerow to hedgerow for less weeds and spray the crap out of it all, just put the cattle out there and let them go wherever they want and trample everything in sight and poop in all the fresh water, but we have grievously streamlined the complex processes of nature with our own oversimplified, lazy thinking.

Wetland ecosystems around the world have been decimated. They include swamps, mangroves, marshes, everglades, estuaries, tidal zones, river deltas, and more. Wetlands rival tropical rainforests for the greatest amount of terrestrial species in the world, with an abundance of birds, mammals, reptiles, amphibians, and plants that can survive nowhere else or depend on the wetlands for their reproductive migrations. Globally, "about 3.9 million square miles of wetlands…, about the size of Canada, [were] destroyed during the twentieth century."[2] In the U.S., half the wetlands have been destroyed and now host almost one-half of all endangered species.[3] "California has lost over 95 percent of its original… 5 million acres of wetlands."[4] Around the Gulf of Mexico, wetlands were a natural barrier against storm surges, but now those protections are mostly gone due to the development of resorts, condominiums, golf courses, and vacation homes, making hurricane damages much worse.

More examples: due to the Newlands Reclamation Project, an irrigation program designed primarily to grow livestock feed in Nevada… the wetlands of Fallon National Wildlife Refuge vanished.[5] The vast marshlands of the Colorado River Delta once covered almost two million acres and were known as one of the great wonders of the natural world, supporting an incredible diversity of life, but it vanished with the building of dams for irrigation and urbanization.[6] The Colorado River no longer even reaches the sea; sadly, the same is true for many rivers of the world. "[T]he ancient marshlands… between the Tigris and Euphrates Rivers" in Iraq, about double the size of the original Florida Everglades, were drained in the 1990s.[7] Russia, known as the most environmentally destructive country in history, got rid of the entire Aral Sea, redirecting two rivers to grow cotton; twenty of twenty-four species of fish went extinct, and the vast blue sea indefinitely became a salt desert with wind storms of toxic chemical dust.[8,9]

In Canada, one acre of forest is clearcut every twelve seconds (in Brazil, the same is every nine seconds), and Canada has only 2.6 percent of its forest officially protected (nearly 400 percent less than Brazil).[10] Such destruction leaves nowhere for animals to go. In Alberta "…the clearcutting proceeds… [there are] no provision for parks, wilderness, fish and wildlife habitat, old-growth forests, biological diversity, watershed management, Indian land claims, subsistence fishing, hunting, and trapping… [just] wood chip production."[11] It is the same throughout all the other provinces of Canada where there are forests.

This means all the original forests of North America, save for comparatively small fragments, will have been destroyed, from Central America to all of Canada, most of which has been converted to farms, cattle ranching, urbanization, or desert. Vanishing are all the wildlife and plant species from forest destruction, destruction so vast it is second only to the ocean's demise. Yes, the forest grows back in some places, but it is replaced with mostly monoculture tree farms or other. It will never be the healthy environment it once was, full of species of all types and free of chemicals, until it has thousands of years to recover. It's not just North America; the rest of the world's original forests, save for what's left of the Amazon and the Russian/Canadian boreal, have already been eliminated by up to 99 percent.

"The U.S. Forest Service is the largest road managing agency in the world, with an official 386,000 miles…" plus an added 54,000 unofficial miles carved out by those who drive four-wheelers, "dune buggies, dirt bikes… ATVs, [and] snowmobiles… [that] tear up the ground, harass wildlife, and act like maniacs…," reveling in the damage they cause.[12] Road construction has destroyed nearly 12 million acres that formerly supported species of all Kingdoms of Life. Logging roads also allow an incredible amount of hunters (and poachers) to have access to killing even more animals; they don't even have to walk or hike, they can just drive around on their ATVs.

The "[t]allgrass prairie, which once swept over tens of millions of acres from Texas to Manitoba, has been largely replaced by farms. Iowa once had 30 million acres of tallgrass prairie, now it has a mere 30,000 acres – a 99.9% loss. Manitoba has only 750 acres out of a former 1.5 million [a 99.995 percent loss]."[13] An earth-shattering amount of native species are gone, from beautiful butterflies and birds to bison and hundreds of grasses and forbs, replaced with an increasing amount of chemicals and GMOs, much of the crops used to feed cattle and make ethanol. Fundamentally gone are the creeks and springs, the fish and frogs... and each year we continue to abandon millions of acres of arable land around the world due to a variety of anthropogenic destruction.

Many believe that technology will save us, yet in an industry which is growing exponentially, the manufacture of a single computer semiconductor chip generates 630 times its weight in waste[14], "and Silicon Valley has twenty-nine toxic Superfund sites – more than any other county in America – and 80 percent of the waste is from electronics manufacturing."[15]

Our natural environment has changed drastically since the human population reached 3 billion in 1960. From then on, nothing has degraded Earth's ecosystems more severely than endless human activity since the beginning of Early Modern Humans 50,000 years ago, a mere one percent of time since our "awakening" as a species.

Corporations have title to vast amounts of land, thinking they have the right to do anything they want with it the same as if it was an acre or two privately owned by a person, but the proven and appalling environmental record of corporate land ownership should disqualify them from even owning land. Plus they have even less regard for human rights every day. To the DC, land ownership means they can destroy it at will for their own benefit without a thought of what it's doing to any other lifeforms, including other humans.

We continue to allow all this to go on as if it's normal and acceptable. We think the Earth can take it forever, will automatically heal it for us, that everything will be all right, or that we'll find the answers to every problem just in time. We think these are isolated incidents that don't accumulate, that our Earth is our mommy, and God or any outside authority is our daddy, and they will both lovingly clean up after our little messes for us, so we don't have to worry. We can just keep on doing what we're doing until we grow up; after all, technology, religion, new discoveries, AI, or even ETs are all there for us – all outside of us, all still like parental figures.

It's not merely coincidental in West Africa, known as perhaps the most environmentally degraded area in the world, that habitat destruction has had the greatest effect on human infectious diseases, as evidenced by the recent, biggest breakout of the Ebola virus in history. This area has suffered from centuries of DC colonialism, currently in the form of MNEs, draining the land of resources while inhabitants there have been forever disregarded, abused, impoverished, and left to live on subsistence agriculture and the poaching of bush meat, including certain primates that have been vulnerable to Ebola for years. The area has been deforested and overgrazed, and irrigation projects dreamt up by the DC "to improve on nature" have backfired. The desert is growing, and lakes and rivers have dried up.

If our species doesn't grow up and become responsible for our own choices and behavior, variations of this scenario will continue to grow in magnitude everywhere. All this is being done by a population of "only" 7.4 billion people....

INVASIVE SPECIES

INVASIVE SPECIES ARE AN enormous cause of the extinction of species, second only to habitat destruction. Invasive species are thought of as weed species that take over where habitats of all types have been at least partially destroyed by human activities, resulting in native species getting crowded out to the point of nonexistence. This bio-invasion threatens over one out of five of the world's endangered vertebrate species and includes oceanic, terrestrial, and fresh water plant and animal species of all kinds. It is a result not only of habitat destruction but of global trade as well. The problem is especially severe, as native ecosystems have no defense against the invading species, which are then able to rampantly expand in their new ranges. In a monetary perspective, "[t]he economic harm caused by the 50,000 non-native invasive plants, animals, and other organisms already in the U.S. is approaching $140 billion per year [as of 2006]," [1] although the problem is the same globally.

From alang alang to zebra mussels, there are too many invasive species on a global scale to list, but there are some of particular note. Many years ago, the prickly pear cactus was intentionally introduced to Queensland, Australia; by 2006, prickly pear covered "over 95,000 square miles in six feet of impenetrable cactus" and is considered Australia's worst weed. [2] Native bunchgrasses in the western United States used to be half as tall as a horse, but now they have all been replaced by such invaders as cheatgrass, pink bull thistle, knapweed, brown curly dock, medusahead, Russian thistle, leafy spurge, and many more; "having already spread over more than 100 million acres of western lands, [they] are invading new areas at the rate of 5,000 acres per day." [3] Highly flammable cheatgrass takeover in particular has been caused by "[o]vergrazing in the sagebrush grasslands of the Great Basin.... 'Cheatgrass is dominant on about... 16.8 million acres... and could eventually invade an additional... 62 million acres.' ...[It] provides little if any food to native wildlife."[4]

Juniper trees, which send out toxins from their roots that prevent other major flora from growing amidst them, have taken over former spruce and pinion pine forests of the southwestern U.S.

Intentional fires set in Southeast Asia "have encouraged the growth of alang-alang (*imperata cylindrical*), a coarse, inedible, fire-tolerant grass that has created wasteland out of millions of acres of land."[5]

Cowbirds have extirpated hundreds of bird species and, in California, are the reason others are now on the endangered species list.

Sea lampreys had always naturally been barred from the inner Great Lakes by Niagara Falls until manmade canals allowed them in; in the following ten years, the annual catch of trout in Lake Huron and Lake Michigan fell by 99.97 percent, "from 8.6 million pounds to 26,000 pounds," destroying the fishing industry there.[6]

"[T]he zebra mussel and Asian clam, have spread in incredible densities through U.S. freshwater ecosystems... (73 percent of three hundred native species are imperiled)."[7]

Domestic pigs became feral in many states, Hawaii being the most notable, destroying ecosystems around the country, and turning oak trees into local ghost species, since the pigs feed on the oak seedlings.

There are estimated to be 100 million feral house cats in the U.S., 2 million just in the state of Wisconsin.[8] This decimates wild bird species and others, yet the cats are innocent; it is humankind at fault, having caused them to be feral through neglect, abuse, and general apathy.

"California's native [beach] grasses have been almost entirely replaced by invaders," such as European beachgrass and iceplant.[9]

"Before humans arrived on the scene, many whole categories of organisms were missing from Hawaii; these included not only rodents, but also amphibians, terrestrial reptiles, and ungulates."[10]

One more example: the water ballasts released from ocean liners traveling the world frees thousands of invasive organisms into foreign waters.

This very short list shows the utter chaos that nature is being put through by various acts of our species, and it has become a severe, global problem that is transforming life on our planet. This further threatens the balance of life as a result of and in combination with all the other negative circumstances humans are causing. It is filling much of the world with opportunistic weed species, which "replace native species, change existing water - or nitrogen - cycling regimes, deprive indigenous animals of their normal diets, introduce new pathogens against which native species have no defenses, and change the genetic makeup of native species by mating with them."[11] This is homogenizing our world like a new Pangaea and drastically decreasing biodiversity, causing thousands of species to go extinct and replaced with weeds.

ICE

THE CRYOSPHERE IS THE scientific name for the major layers of ice that form on the surface of our Earth and its oceans. Glaciers once covered various parts of Gondwana as it swirled over and around the South Pole before it stretched north and meshed with the other remaining continents to form Pangaea, eventually leaving Antarctica where it is today. Although glaciers were forming on Antarctica up to 30 million years ago, it took almost 28 million years more for a permanent ice cap to form on the Arctic Ocean. Our current ice age began circa 2.4 million BP, about 280 million years after our Earth had been ice-free. After another half million years, ice sheets up to over two miles thick covered much of the Northern Hemisphere, melting and returning about twenty times until our current 11,500-year Holocene geological epoch began, the glaciers having taken 7,000 years to melt from their previous maximum range. The previous interglacial period of warmth was about 100,000 years ago.

Our ice is now melting at a historically unprecedented rate, in less than a miniscule 2.4 *millionths* of one GY (100 years). The Arctic ice cap is diminishing at a rate that fluctuates between 3 to 9 percent per year and has lost an area of over 400,000 square miles since 1980 (800 by 500 miles or four times the size of Colorado). Normally, from summer to winter, the Arctic sea ice doubles in size, gets to about three feet in thickness, and has permanent pack ice of about sixteen feet; however, these thicknesses are decreasing by up to 40 percent. "In the Arctic, perennial sea ice covers just half the area it did thirty years ago, and thirty years from now, it may well be gone entirely."[1] The heat erosion of ice is happening to the rest of the cryosphere around the world as well - at the southern and marginal regions of Greenland, for example - and "[a]lready the Alps have lost a quarter of their ice [2006]. Alpine melting in 2002, during the warmest-ever summer in the Northern Hemisphere, contributed to unprecedented floods in central Europe."[2]

The Himalayas, often referred to as "the Third Pole," consist of an ice pack that contains 46,298 glaciers, and they are melting rapidly. The rivers they supply bring water to over a billion people in Asia, and without such ice the rivers will become seasonal. "The Ganges is now in such serious decline that it is considered among the 10 most endangered rivers of the world."[3]

Antarctica's ice melts more than the regions of the North because its summers are warmer due to our current time within the 25,700-year precession of the equinox. Antarctica's annual sea-ice cover melts almost entirely every summer; here, the average annual temperature has risen by 11°F, and the Pine Island Glacier, for example, is in irretrievable meltdown, adding 46 billion tons of fresh water to the ocean and melting back at an incredible 2.5 miles per year. Irretrievable meltdown happens when the ice that has melted creates a lubricant of water underneath the glacier, making it move faster and calve larger and more. The albedo also lowers.

The albedo is a measurement of our Earth's surface reflection of the Sun and the heat which is absorbed from it. The measurement is from 1 for complete reflection to 0 for complete absorption, the average for the entire Earth being 0.33 or two-thirds absorption of the Sun's heat. Clouds can reflect up to 100 percent and ice about 80 percent; on the other hand, the ocean absorbs up to 90 percent of solar radiation, an albedo of 0.10. In a property known as "latent heat," it takes eighty-one times as much heat to melt ice as it does to raise the same amount of water by one degree.[4] A lot of heat is needed in order to melt a glacier, but as it does, the surrounding albedo lowers and the area absorbs more heat, creating a positive feedback loop which intensifies the situation to the point of irreversibility.

If anyone still wants to deny that climate change is happening, the movie, *Chasing Ice*[5], very well documents the cryosphere melting with time-lapse photography over a period of years. In their rapid retreat, glaciers are deflating like gigantic air-bed mattresses. The movie also points out how the professional deniers only focus on the four glaciers in Alaska that are growing rather than the over 500 that are melting there.

It's still uncertain how much the ocean levels may rise. The study of earth sciences has allowed us to realize that ocean levels have previously gone up and down by as much as 300 feet from our world with no ice to our world with maximum glaciation. Research also reveals that the diminishing cryosphere is directly related to the increased levels of CO_2 human activity has been adding to the atmosphere. This is proven from hundreds of core ice samples that have been studied, some of which even have ancient air bubbles trapped inside the ice, making it a very exact science. The maximum depth of ice to test goes back 800,000 years.

DESERTIFICATION

"'Desert-makers' is truly as appropriate a
title for humans as 'tool-users.'"[1]

T HE DC BEGAN TAKING with the first explorers. Nature was in
such great abundance they thought the resources would never end,
and so began the age of great waste and frivolity. Hunting for food was
first. The Great Auks of the North Atlantic were pushed into extinction,
the last three strangled by hand by two Scottish men whose names were
even known for this conscious act.[2] As with the well-known extinction of
the dodo birds, big as turkeys, as easy to catch as the auks, other cultures
did the same: the Hawaiian royalty donning their yellow-feathered
capes that required tens of thousands of feathers to make, each *mamo*
or *oo* bird providing only two feathers each; the great moa birds of
New Zealand before the Maori; the elephant bird of Madagascar; and
though the DC should know better by these examples, it continues, as
with the Bluefin tuna and many other species today.

The trappers followed the explorers, such as in North America and
Far East Russia, selling furs to make money. It was a natural, time-
honored means of providing warmth for millennia and especially took
hold during the sixty-year miniature ice age in Europe and Northern
Asia beginning in 1645. By the 1800s it became an economic industry
that has never stopped; fur fads continue to this day, a means for the
clueless, uncaring, and narcissistic who unconsciously follow obsolete
customs.

Once the inroads were made by the trappers, the loggers came in
to make their fortunes in decimating the forests as fast as they could,

to grab as much of the profits before anyone else, under the guise of providing for the betterment of society. Then they simply walked away from the ravishment they caused, moving on to do it somewhere else without so much as planting a single replacement tree for nearly 500 hundred years.

When it was safe enough to risk it, the settlers then followed, clearing the rest of the forests of their remaining trees, other native plants, and wild animals in order to establish their homesteads, plantations, and farms, extirpating or causing extinction to thousands of species. Some of these forests never returned, making the soil warm up and dry out, dropping the water table, and radically changing ecosystems, especially where cattle ranches were established. Once cattle and other domestic ruminants took the last that could be taken from the land, the land was abandoned as a dried-up desert, devoid of retaining water, unable to support anything but the strongest of inedible weed species. With the protective forests and thick prairie vegetation gone, the soil exposed and dry, water and wind have been flooding and blowing soil away ever since.

If the anthropogenic wasteland that remains is found to have any valuable minerals in it, the mining companies complete the exploitation of the Earth, including leveling whole mountains for coal. Other MNEs are now even fracking for natural gas and puncturing ever-deeper layers of rock for oil, oil which is under nearly uncontrollable pressure, as was discovered by the British Petroleum oil spill in the Gulf of Mexico in 2010.

The explorers and trappers got rid of most of the animals, the loggers and settlers got rid of the forests, the cattle ranchers and farmers got rid of the native plants and eroded the soil away, and the miners finished the job of total depletion, total extraction, total exploitation, all to gain personal wealth over others with the excuse it was helping mankind. Even the water tables and underground aquifers are disappearing or being chemically contaminated. Now many dams are approaching deadpool, without enough water to even flow past the dam or turn the electric generators, and more and more rivers don't even reach the sea, even the mighty Nile.

Positive feedback loops make the once-healthy and balanced Hadley Cell turn for the worse: its wet aspect worsens tropical floods and storms, intensifying the erosion of logged-off tropical rainforests, and its dry aspect spreads the deserts in every direction. Storms of every kind grow more disastrous. Even polar caps melt, adding to the planetary positive feedback loops.

"We are turning prairies, rain forests, and marshes into deserts and wastelands. In taking over more and more of nature for human life, we are, ironically, destroying more and more of the quality of human life."[3]

> "One consequence of the land-use cascade [the process by which wildlands are appropriated for human use and then exploited until exhaustion brings about abandonment] is certain: wasteland will continue to accumulate through the coming decades. Additions to wasteland will come from many sources: erosion of existing cropland and pastures, soil exhaustion, salinization, aquifer depletion, desertification, and, increasingly, urbanization. Wasteland is accumulating far more rapidly than it is being restored to productive use, not because of a lack of scientific know-how but because there are few economic or political incentives for rehabilitating wasteland...."[4]

"Since 1950, an area worldwide equivalent to the combined national territories of China and India has been abandoned to wasteland. If one also counts as wasteland those areas taken over since 1950 by industry, urban expansion, and road building, the total is substantially larger."[5] "In the United States, 'at least moderate desertification... [affects] 90 percent of the arid lands,' and around the world, desertification has hit hardest in cattle-producing regions, such as the western United States, Central and South America, Australia, and sub-Saharan Africa. The grazing pressure exerted on the world's grasslands by livestock is about ten times higher than what such landscapes would naturally support in the way of native herbivores."[6]

"Nearly 2.5 million... acres of cropland, an area larger than Yellowstone National Park, is being abandoned annually in the United States because of erosion and other agriculture-induced degradation."[7]

"Half the topsoil of the Midwestern farm belt of the U.S. is thought to have washed into the Gulf of Mexico since the land was first plowed a century and a half ago. A typical farm in Missouri loses twenty tons of soil per hectare per year, whereas the rate of new soil formation is only one ton per hectare per year."[8] We're losing topsoil twenty times faster than it can be replaced.

"From 1970, the proportion of the Earth's land surface suffering very dry conditions rose from 15% to 30% at the start of the twenty-first century."[9]

The DC ideology began in the lands around the Mediterranean where forests were logged repetitively for millennia. With the continual loss of trees which secured and regulated the water table, transpired into the air, provided shade, and slowed the winds, vast areas of land dried out and became desertified. The difference now is this same process has become planetary in scale.

OCEAN

SINCE WE ARE TERRESTRIAL creatures, we are unaware and forget how truly our planet is ruled by oceans, oceans that cover 71 percent of our Earth's surface, a liquid planet contained by its own gravity, floating, rotating, and orbiting in space, ruled by cosmic forces and planetary dynamics that are immensely powerful, entirely primordial, and transcendent of humankind. More than the surface cover, the ocean volume, with an average depth of 2.37 miles, consists of 330 million cubic miles, containing 99 percent more viable room for life than land. As life originated in the ocean, to this day, of forty-four phyla extant in the ocean, thirty-two are animal; whereas, of twenty-eight phyla extant on land, only twelve are animal.[1] Our weather is generated all from the ocean, made more turbulent once it hits land by mountains and thermals, but the terrestrial 29 percent of Earth can only receive whatever weather it gets from the ocean. Whatever happens to the ocean profoundly affects all life on Earth.

The oceans comprise 97 percent of the water on Earth, with an average salinity of 3.5 percent, caused by the weathering of terrestrial minerals from rains brought in by the ocean, assisted by the fungi and lichens that slowly decompose rock. Each year, 35 billion acre-feet of water evaporates from the ocean then rains down, filling rivers that deliver 4 billion tons per year of dissolved chloride, sulfate, magnesium, calcium, and potassium salts back to the ocean. Furthermore, the oceans are saltier in the subtropics (20° to 30° north or south) than in the tropics due to the massive rains in the tropics from the Hadley Cell. Evaporation causes salts to condense in the ocean, but the salinity balance is maintained by the continual settling of salts as sediment on the bottom of the ocean, eventually submerged under continental plates by tectonic subduction, and becoming rock again through subsequent uplift and volcanoes over eons. It is probable the same percentage of salts are in our blood as the oceans had when the first amphibians ventured onto land, indicative of our evolutionary, oceanic origin, our direct genetic ascendancy, and our intrinsic oneness with our Earth.

The oceans contain much more carbon dioxide than the atmosphere because of density, and the ocean absorbs about 30 percent of atmospheric CO_2. This means the oceans have taken in about 100 billion extra tons of carbon since 1800, are now absorbing 2.5 billion tons per year, have become 30 percent more acidic, and will reach 150 percent by 2100, a rate understood to be unprecedented in Earth's history.[2] The acidic levels we are already approaching have not existed for more than 20 million years and will take several thousand years to rebalance on their own. Normally, the average pH of the ocean is 8.15, which is slightly alkaline. As the ocean continues to become less alkaline from CO_2 absorption, without humans changing "business as usual" oceanic creatures will be unable to form shells, and up to two-thirds of ocean organisms will cease to exist, as the areas around ocean vents with the more-acidic chemistry already show. The Arctic is the only ocean not directly affected by the thermohaline cycle, making acidification worse in the northern seas, which affects the abundance of plankton there more than the hole in the ozone does over the Antarctic. The thermohaline "conveyor belt" circulates deep ocean waters on a thousand-year cycle, regulating the ocean's chemistry, temperature, and distribution of nutrients, nutrients which upwell along continental margins and allow coral reefs and kelp forests to form, further developing the abundance of life they support.

Our oceans consist of two layers, the surface and the deep. The surface layer extends to depths of 200 feet, is turbulent, well-mixed, and rapidly absorbs CO_2 from the atmosphere. The deep layer extends from the surface layer to the ocean floor and contains much more carbon. The more carbon the ocean absorbs the deeper and more long-lasting its turnover rate is, which is the time it takes for the effect of CO_2 to be removed from the environment. This rate is historically from eight years at the surface to a thousand years at the bottom; currently, the surge of extra carbon added to the ocean by humans has pushed the minimum carbon turnover rate to sixty years rather than eight, and this number will continue to grow for centuries even if we stop adding carbon to the atmosphere now.[3]

Our oceans are warmer now than they have been for almost 500,000 years, as they absorb heat a thousand times more than the atmosphere; likewise, the ocean absorbs more heat than land, as it has a much higher absorption rate (lower albedo) than most land surfaces. As more reflective (high-albedo) sea-ice melts, the ocean absorbs even more heat, affecting its own chemistry and all life on Earth.

Thirteen tons of garbage is thrown overboard from ships at sea every day. Every year 200 billion pounds of plastic are made, 2 billion pounds of which are tossed into the ocean.[4] Garbage from our species accumulates in an eddy double the size of Texas in the Pacific Ocean north of Hawaii, and others form in the Atlantic and Indian Oceans. Tons of garbage wash up on beaches all over the world; microscopic plastic seeps into the soil and food chain; radioactive waste is dumped into the Arctic Ocean and others, and the Pacific receives untold amounts of leaked radioactivity from Fukushima, unchecked now since March, 2011. Multimillion-gallon oil spills happen regularly. Military tests pummel the oceans' depths with mega-pulses of sound. Every year this results in millions of additional seabirds, fish, mammals, and other ocean life dying from human causation. Furthermore, though they are now over 95 percent gone, the ocean's top predators contain various amounts of PCBs, mercury, and radioactivity, as these and other poisons bioaccumulate up the food chain.

We are causing our oceans to become a deserted, toxic void.

FISHERIES

BEFORE 1900, THE TOTAL annual catch of all species of fish never exceeded 10 million tons. With twentieth-century technology, the global fish harvest rose to 20 million tons by 1950. It doubled in just ten years to 40 million tons. In twenty more years (1980) the global fish catch doubled again to 80 million tons.[1] Forty years ago, it was estimated that 450 million tons of fish was available for harvest from the ocean[2]; however, the estimate of fish available for harvest has fallen by 80 percent to 90 million tons due to continuous overfishing and waste. This is the same amount humans now catch each year from Earth's vast oceans, which means we're catching as much harvestable fish as is available. Guesswork is no longer involved: our development of high technology, including the use of satellites and a command post of on-board computerization, allows us to pinpoint where the fish are.

Industrial fishing began in 1950. Since then, the fishing industry, privatized and taken over by giant corporations with enormous factory ships, receives about $25 billion in annual government subsidies. According to the UN, the industry has twice as large a global fleet as the oceans can support. The industry is allowed to fish in 99 percent of the ocean; has devastated the shallower waters of the oceans with bottom trawling, which is comparable to clearcutting the forests on land; and has reduced the big fish populations, such as Bluefin tuna, swordfish, halibut, and marlin, by over 95 percent. Furthermore, it has trawling nets with openings large enough to encompass thirteen 747 jumbo jets and places fishing lines 100 miles long with 2,000 hooks each, setting 1.4 billion hooks per year, which entails using 13.2 million miles of line, enough to encircle the Earth 550 times.[3] That's enough line to extend fifty-five times further away from the Earth than the Moon each year.

Overfishing the oceans has been going on for decades. Stocks of wild fish have been in steady, alarming decline. By 2003, one-third of all edible fish stocks were gone, collapsed to the point of non-rejuvenation, no longer able to restore their numbers in their massively altered circumstances. Overall, fish have become reduced in size as a result of human selection forever harvesting the largest; over time, only the genetically smaller specimens remain to propagate the species. Cod fisheries of the Grand Banks in the Northeastern United States and Newfoundland, Canada, are the most notable example of this: up until the 1960s, the average cod weighed twenty pounds, but in thirty years of industrial overfishing – a time when the human population nearly doubled - this area's cod fisheries, including haddock and flounder, collapsed to the point of no return, reduced by 99 percent. The few surviving cod now average only three pounds. Every year, humans need more fish similar to cod - such as sea bass, tilapia, and pollock - than the Grand Banks provided at its highest recorded levels. Of these cod substitutes, pollock is the only one still substantially caught in the wild at 2 billion pounds per year. Most other fish are now raised on commercial fish farms.

Commercially farmed fish include salmon, Bluefin tuna, tilapia, and sea bass, among others. Salmon and tuna are carnivores. It takes up to five pounds of fish consumed by humans - such as sardines, anchovies, and others - to raise one pound of salmon, and it requires up to twenty pounds of similar fish to produce one pound of tuna.[4] Thus, forty percent of the wild fish catch (36 million tons) goes to feeding farm fish.[5] In other words, people eating salmon and tuna, whether farmed or wild, are at this point depriving less fortunate people of other edible fish; additionally, wild salmon or tuna have ever-increasing levels of PCBs, mercury, and radioactivity in them. Sea bass and tilapia are herbivores, cost much less to raise, and do not compete with human consumption of other fish. Tilapia production has grown by about 330 percent over the past twenty-five years, mostly in China where two-thirds of the world's fish farms exist.

Salmon have been in existence for 50 million years, but the only wild salmon left in the world come from the wilds of Alaska and Russia's Far East. Three-quarters of Pacific salmon are now threatened with extinction; wild salmon returning to their spawning grounds all along the Pacific coast from Alaska to California have been reduced by about 98 percent and extirpated from at least forty rivers. "California closed its salmon fishery completely in 2008 for the first time in history."[6] The Pacific Ocean used to contain easily over 100 million salmon. The Columbia River alone used to have 15 million salmon return to it to spawn all the way into tributaries in Idaho and Canada; however, the architects of the Grand Coulee Dam never had fish ladders built, permanently cutting off 3,000 miles of salmon spawning grounds. Now many wild salmon species of the Columbia are extinct, and few of the grandest of all - the 40-50 pound king salmon - barely show up even in Alaska. Thus the rise of industrial fish farming began.

Whether they're labeled from Canada, Chile, Scotland, Ireland, Norway, or as "Atlantic salmon," all these salmon are farmed.[7] Chile is the second-largest producer of salmon in the world. Farm salmon have been so genetically modified (GM) that scientists refer to them as *Salmo domesticus,* and they regularly escape into the wild, contaminating wild salmon with their GM DNA, making even wild salmon evermore incapable of surviving without human assistance. Substances have to be introduced into farmed salmon to make their meat look pink rather than grey; they are, in a sense, artificial salmon.

As coastal fisheries suffered decades of exploitation and severe degradation, humans developed the technology and necessity to expand out to the more demanding task of fishing for pelagic fish in the open ocean. International agreements concerning tuna fishing proved futile, and a high seas free-for-all has ensued; what regulations exist are nearly impossible to enforce on the open oceans, and laws allow for fishing across international borders.

Three species of Bluefin travel the entire expanse of the Atlantic, Pacific, and Southern Oceans, breeding in the Mediterranean Sea, the Gulf of Mexico, and near Japan. Bluefin tuna are becoming so rare that in January 2013, one 400 pound Bluefin was bought by a Japanese broker (where 90 percent of the tuna brokerages are) for $1.3 million dollars; admittedly, it was partly a marketing ploy, but the tuna still drew a price of $3,500 per pound. In the real-world market, "[a] single wild Bluefin tuna will often sell for more than ten thousand dollars."[8] With so much value put on Bluefins, it also encourages the Ghost Demand Syndrome (when an endangered species' value goes up, it becomes more sought after before it becomes extinct, further ensuring and quickening its extinction).

Bluefin tuna are like no other fish on Earth. These iridescent fish are warm-blooded, have retractable fins to streamline themselves for swimming across entire oceans at up to fifty miles per hour, and at times move their tails as fast as hummingbird wings. These highly specialized apex predators can grow past fourteen feet long and up to 1500 pounds.

Bluefins are slow to regenerate, taking seven years to reach sexual maturity and even longer to become the primary breeders, those over 500 hundred pounds, the ones that were most exploited by industrial fisheries. Since Bluefins are now so rare, they are being commercially farmed. Loopholes in the law allow juveniles to be caught in the wild and then hauled in pens to harbors to be fattened and sold before they ever have a chance to reproduce.[9] This is causing the "candle to be burned at both ends" of the tuna population.

Just 3 percent of the world's Bluefin tuna population remains. Some estimate there are only a few thousand left in each of the three ocean groups. Ninety percent of them are less than two years old and less than three feet long. Bluefins once numbered in the millions; now the most extraordinary fish on Earth are in grave danger of becoming *extinct*.

The only lifeforms of the high seas protected by the laws of humans are whales; it's far past the time to extend similar laws to Bluefin tuna. The Mediterranean Sea is practically void of Bluefins, as the Egyptian High Aswan Dam has indirectly destroyed the Bluefins' food source of sardines; the Gulf of Mexico is lethally toxic to apex predators since the BP oil spill; and the waters around Japan become more radioactive every day from Fukushima - the three main Bluefin breeding areas. At this point, catching, buying, and selling Bluefin tuna on every level should be fully banned and illegal. This precious species and food resource must be allowed to rejuvenate its population immediately; otherwise, it passes into *extinction*.

Combined with all the other modes of fishing, over 7 million tons of sea life per year is completely wasted, called bycatch. Annual bycatch includes hundreds of thousands of porpoises and whales, 100,000 albatross, 200,000 tons of sharks, untold numbers of endangered sea turtles, and on and on. For example, it takes up to ten pounds of bycatch, including endangered species, to produce one pound of wild shrimp. Bycatch is hooked, gaffed, beaten, stabbed, crushed, eviscerated, maimed, mangled, and thrown away all day and night, every day. Additionally, continental-shelf habitat annually scraped by bottom trawling equals roughly one-tenth the world's land surface[10], and it's similar to leveling a forest in damage.

"The United Nations' Food and Agriculture Organization (FAO) concluded that approximately 75% of the world's fish stocks are fully exploited, overexploited, or depleted, while IUU [illegal, unreported, and unregulated] fishing may amount to a third of total world catches."[11] Illegal fishing in 2011 amounted to $16.5 billion in revenue. As on land, the objective is to maximize humans' extraction of animal resources, which at best always keeps their numbers at minimum levels and marginalizes unvalued lifeforms as economic nuisances. Plans of restoring wild species' numbers to their historically healthy balance are never even pondered because it's impossible anyway, so far.

The Amur River is one of only three major rivers in the world that remains free of any dams. The Lena, in Russia, and the Amazon are the others. The Amur is the tenth longest river in the world and forms the border of Russia and Mongolia. It has "180 species from 79 genera, 23 families, and 10 orders from a bizarre blend... of species that occur nowhere else on earth... [many] from the Pleistocene and Pliocene epochs."[12] Russia abolished its equivalent of the EPA and Forest Service in 2000, so with the Amur in economic impoverishment, tons of critically endangered species are being illegally caught and sold on the open market, as any means to survive rule the day.[13]

Of the 330 species of sharks, dozens are approaching extinction. In 2013, 100 million sharks were killed at sea, their fins hacked off and their agonized bodies left to drift away, salt water searing their wounds. In Asia, "...shark fin soup can sell for as much as $200 US a bowl."[14] In areas around the world where sharks used to breed, such as the Gulf of California, the waters are now deserted. It takes a long time for sharks to reproduce. As on land, without apex predators, the oceanic ecosystems are drastically thrown out of balance, resulting in a deluge of trophic cascades.

In a classic example of a trophic cascade, in 1999,

> [T]he *New York Times*... reported that overfishing, along with climatic changes, has led to a crash in the herring and pollack [sic] populations. This, in turn, has led to a steep decline in the populations of sea lions and seals who feed on them. Sea lions and seals are an important prey species of orcas, so now the orcas had to hunt other prey, including sea otters, in much shallower water than they usually frequented. This created a dramatic 90 percent decline in the numbers of sea otters. In some places they vanished altogether. The kelp forests, home to the sea otters, were now affected by a sudden massive increase in sea urchins, a favorite food of the sea otter and the main predator of the kelp. The devastation of the kelp forests affected the multitude of marine life that they normally support. Mussels, fish, ducks, gulls, and bald eagles were all affected, as were the already diminished populations of sea otters, sea lions, and orcas."[15]

In the big picture, humankind began fishing on land in rivers and lakes and shallow coastal waters. As boat technology expanded, so did our ability to fish deeper waters further from shore. By 1950 we were so capable of pelagic fishing that sixty years later the entire ocean is completely overfished. We've cleaned out the ocean enough that we've had to start raising our own fish on fish farms, leaving increasingly less food for other species in the oceanic wild. About half of Earth's vertebrate species live in the ocean, and half of those surveyed are facing some threat of extinction[16], and, since 1970, there is over 50 percent less of life in the ocean.

ANIMALS

T HE KINGDOM OF ANIMALIA has more varied forms of life than any other kingdom, even though 80 percent of all animals are nematode worms.[1] The other 20 percent is composed of insects, corals, clams, birds, jellyfish, reptiles, mammals, amphibians, and so on. All thirty-two animal phyla evolved in the ocean, most of them roughly 540 million years ago. There are about 45,000 species of vertebrates, and all of them have gill slits in the pharynx or throat at some stage of their lifecycle, whether as embryos or adults, revealing their marine ancestry[2], including humans. From the ocean, then onto land, animals are the only organisms on Earth that have been able to inhabit the atmosphere; no other kingdoms of life do so except in their methods of dispersal (spores, seeds, etc.). Animals are the only lifeforms that fly, and some spend their entire lifecycles in the atmosphere. Since flying reptiles (dinosaurs) are extinct, only birds and certain insects have taken wing, with the one addition of bats, the only form of flying mammals.[3] Bats are from a very old mammalian ancestry and include about 20 percent of all mammals, although their numbers have been decimated recently in North America, and 26 percent of bat species face extinction.

Approximately a mere 5,500 species of mammals exist, a very small percent of the animal kingdom. The number of mammals is 40 percent less than the number of bird species, another small percentage of animals. Furthermore, 40 percent of mammal species are rodents, leaving only about 3,300 other mammalian species remaining. More mammals exist towards the equator, of course, as is true for every kingdom of life; unfortunately for non-human mammals, it's also where most humans and poverty exists.

Vietnam is a poignant example of what happens to animals in poor, tropical countries around the world. Vietnam extends from 8^0 to 23.5^0 north latitude with a coastline longer than the Pacific Coast of the continental U.S.; however, in size Vietnam and the state of New Mexico are nearly equal. Moreover, Vietnam's population is over 93.4 million people compared to New Mexico's 2.1 million. As people look for any way to survive in an impoverished country, the past forty years have caused an alarming and growing number of large animal species to go extinct in Vietnam, mainly due to illegal wildlife trafficking.[4]

The black market industry in the illegal capture, killing, and trade of wild and endangered species approaches $20 billion in annual profits; it is so rampant and established that the international trade is third only behind illegal drug and weapons trafficking. This holds true for all equatorial countries around the world that export mostly to more northern countries. Moreover, many of the same smuggling routes are shared by the criminal drug and wildlife trades, and illegal drugs are even hidden within legal shipments of wildlife. One-third of cocaine captured in the United States is accompanied with legal shipments of wildlife.[5]

Another example: up to 38 million animals are caught every year in Brazil, but only ten percent of them survive the process of capture and transport[6]; furthermore, once these animals reach their destinations of ignorant demand, they do not survive long. "According to the entry on pets in the *Encyclopedia of Biological Invasions*, every year more non-indigenous species of mammals, birds, amphibians, turtles, lizards, and snakes are brought into the U.S. than the country has native species of these groups."[7] It is a profitable business at the end of the supply chain; the values rise as high as 6,000 percent from the price paid to the original captors. The illegal trade of wildlife is bringing animals to the point of extinction and causing the collapse of ecosystems.[8]

Regarding nonhuman mammalia, "[n]early half of all primates, the order that includes [lemurs], monkeys, and apes, are threatened with extinction,"[9] many critically. Alarmingly, the average of mammals in all the other non-rodent mammalian orders facing extinction is approaching one-third of the 3,300 species that exist.

Animal fur has been in high demand throughout human history. With the rise of man-made textiles, fur no longer became essential for most people to stay warm. A popular movement arose about forty years ago to end the unnecessary and inhumane killing of animals for their fur, worn mostly as a fashion product for those now considered ignorant and of low status, and the movement had an effect. However, since 2000 the fur industry has been experiencing a revival; sales of fur coats were up by 20 percent from the year before to a total of $1.68 billion. That means "in one year, between 7 to 8 million animals were killed for their fur just in the United States alone. And worldwide... 28 million animals were farmed, and 7.6 million trapped, for their fur."[10]

It's not just fur and mammary glands that make mammals unique in all the Kingdoms of Life; even more remarkable is the development only in mammals of a limbic brain. The limbic brain is what creates and allows emotions to be felt. There is no debate on whether mammals feel emotion; many mammals show emotions that are very likely identical to those we know such as fear, joy, rage, love, sadness, despair, grief, compassion, and many more intricate.[11] Knowing this, DIC members still continue to treat mammals with horrific and torturous abuse, seeing them primarily for sources of or obstacles to money, for research, for transferring human pathologies onto, or for feeding extravagant and unsavory human egos.

In North America before 1600, tens of millions of beavers created vast wetlands and riparian forests; tens of millions of bison, antelope, and elk roamed the continent, as did mountain lions and gray wolves... [and] woodland caribou...,[12] yet now their populations are all from 95 to 99 percent diminished. Animal populations are in trouble all over the world, even from infectious, parasitic diseases, such as toxoplasmosis from domestic felines causing half of sea otter deaths from humans flushing cat feces down toilets.[13]

Most seem to erroneously believe everything will be fine because we will at least be able to maintain the last members of any species somewhere in zoos, but zoos are where animals become completely dependent on humans for everything, including reproduction. Zoos are where mammals literally go crazy, become despondent, and die after they act out strange psychological/emotional pathologies, including "[c]annibalism, self-mutilation, hypersexuality, eating disorders, and a whole range of pathetic behaviors... even in 'progressive' institutions...."[14] Zoos are often the last stage upon which endangered animals are placed before they go extinct. Parks, reserves, and refuges will not save them either, which are all compromised and fragmented, as we are already seeing in the case of rhinoceroses and many others.

In 2014, the United Nations Global Diversity Assessment counted 36 percent out of over 52,000 assessed animal species that face some threat of extinction. Lost in our own world of technological gadgets, economics, and politics, animals have faded far into the world of the forgotten, irrelevant, and abandoned, as humans rush headlong into causing an ineffably tragic mass extinction of species, without a care, while they live frivolous modern lives, fooling themselves that they're accomplishing anything of worth.

RELIGION, PART I

RELIGION STARTED WITH WORSHIP of the Sun. The Sun symbolized God, the bringer of light, the slayer of dark, and the abundance of food security, especially once agriculture began 10,000 years ago, that 2.4 ten-thousandths of one GY out of the 19.17 GYs of Earth's existence. Sun worship was most apparent with the agricultural empire of Egypt with its devotion to Ra, the Sun God, and Egyptian religious beliefs established the template for the religions of Western Civilization (WC).

Over 5,000 years ago Horus represented Ra on Earth; he was the earthly Sun God and conceived without a father (i.e. a virgin birth) and born from his mother, Isis, a name which translated means life and light. Egyptian religion also included belief in the afterlife and eternal life, the holy trinity, final judgment, the Madonna and child (Isis), circumcision, the ark, the great flood (of Gilgamesh), baptism, and more. This is the same mythology transposed later onto other Western religions.

No one knows how the symbol of the Ankh came about, but Egypt is placed around 26° north latitude, and from Egypt's viewpoint 5,000 years ago, the Southern Cross constellation would've been on the horizon at either dusk or dawn and conjunct with the Sun where the Sun set or rose at certain times of the year. (The Cross has since moved south of Egypt's horizon due to the precession of the equinox). This alignment would be very important to an agricultural society based on astronomical observation as Egypt was. The conjunction of the Sun and the Cross could easily have been portrayed as the symbol of the Ankh. There are other connotations with this alignment such as the metaphorical birth, crucifixion, and resurrection stories of the Sun God. One thing we do know about the Ankh is the Pagan cross and the

Christian cross both originated from this symbol: Egypt has the Sun on top of the Cross, whereas the Christian and Pagan cultures symbolically place the Sun in the middle of the Cross. The Sun is also represented as the "aura" around Christ's head like the Ankh.

Throughout history, other cultures of WC used versions of the Egyptian mythology based on observations of the Sun and nature, later branded as Paganism, the devotion to and respect for nature: Attis of Greece, Mithra of Persia, Dionysus of Greece, the gods of Rome, even Buddha and Krishna of India, Quetzalcoatl in the Americas, and more. All have versions and variations of either being born of virgins or performing miracles, being resurrected, having twelve disciples they traveled around with (as the Sun has twelve constellations it travels through), and many other common similarities. Once again in the Middle East at AD (Anno Domini, Latin for "in the year of the Lord"), another (Son) God was said to have been born of a virgin, performed miracles, traveled around with twelve disciples, was crucified, was dead three days, resurrected, and more.

None of these stories, however, can be proven as fact; all is myth and legend and hearsay. This only points out that the Jesus story is one of a long line of similar stories, the others of which are mostly ignored or forgotten (except by the likes of Joseph Campbell[1]). Many scholars of these historic myths refer to the bible as a hybrid because of how much the bible borrowed from those previous cultures, cultures that based their religious mythologies on astronomy and worship of the Sun. No historians of the time even mention Jesus, and no one has ever proven that he ever existed. Conveniently, we're told that the lack of proof is purposeful in order to test our faith, which is a controlling denial tactic used by the DC.

Seen as the myths and metaphors biblical stories were meant to be, there's no problem; it's when the stories are taken as literal truth and every word is adamantly clung to that all the combativeness begins. Believing in a myth can be useful, and it's part of the core of a culture, but it can also keep people from evolving and becoming free with deeper or actual truth. Religious myths support core cultural beliefs and have led to fundamental mind control. Different religious theories have tried to make sense of the world for thousands of years, but "all they are is theories, theories built on myths. Zoroastrianism – some live in the light, some live in the dark; the polytheism of Greece and Rome that the gods were just playing with us; Buddhism, that peace can only be had by giving up all desire;" and the current theory that Adam and Eve caused all of humankind to fall from grace, all then "to live by the sweat of [the] brow, miserable, alienated from God, and prone to sin and… rewarded for good works [only] after death."[2]

The Judaic calendar marks this year as 5776 BP, so Judaism can date itself back to 3,760 years before Christ. The Judaic Torah, the first five books of the Old Testament, was written about 1445 BC, so it's almost 3,500 years old. With the addition of the New Testament, the Christian bible was completed about 100 years AD or about 1,900 years ago. The Judeo-Christian bible took a process of at least 1,500 years to write.

Islam came into being with Mohammad in 622 AD, over 500 years after the last books of the Judeo-Christian bible were written. Mohammad was not able to read or write, so the Quran was written from memory after he died - as was the case with the books of the New Testament after Christ died. Judaism, Christianity, and Islam all admittedly come from the same grandfather; they're all from the same family of Abraham. According to myth, Abraham had two sons, Isaac and Ishmael. The Judaic religion (and thus Christian) comes from Isaac, and the Islamic religion comes from Ishmael. Islam further breaks down into the Sunni and Shi'ite factions who still fight over who the rightful inheritors are to lead the Muslim nation after Muhammad's death. However, that's nothing compared to how much Christianity has fought and fragmented from true Christians to Catholics to Protestants, Lutherans, Methodists, Mormons, Episcopalians, Baptists, Calvinists, Adventists, Presbyterians, Angelicans, Pentecostals, and many more. Christianity is the only religion of all the religions of the world that believes Jesus was the Messiah, the Son of God, and this belief was forged into the DC like a blacksmith beating it into a form of steel for centuries. Nonetheless, all religions claim to be the one true religion, the one that's got it right.

However, the point is not whether *any* religion or religious story is true or not or who's got it right. The point is, fundamentally, every religion on Earth teaches that the Earth belongs to man, not that man belongs to the Earth, and that's the central flaw of them all.

This distinction is the primary difference between the DC and the cultures of indigenous people: the indigenous have always seen themselves as part of the Earth, in no way seeing themselves as separate, and know how to live with the Earth. In opposition, the DC has theologically and ideologically separated itself from nature with core beliefs and does not know how to live in harmony with the Earth. For instance, the Abrahamic religions maintain that 'spirit' is separate from nature and rules over it from without.[3] That's why the DC, in its complete arrogance and ignorance, has been destroying our Earth. The DC is cut off and mutated away from the rest of nature just like cancer cells are cut off and mutated away from their genetic memory in a healthy body, and the untreated, disconnected cells eventually kill their host as is happening.

The three Abrahamic religions have been a very dysfunctional family even though they all share the same basic ideology, which is the foundation of Western Civilization. They've been murdering one another for millennia, all the way back to the original son, Cain, killing his own brother, Abel. All three of these religions believe in the creation story of Adam and Eve, which was written by the Hebrews, the Semites, simply a tribe at the time. Every tribe on Earth has or had their own creation story of how their people originated on Earth. Ask any Native American tribe. The difference is, the Hebrews wrote it down - as did the Egyptians, Christians, and Muslims - and doing so made it carry more importance, gave it more legitimacy as being true at the time in a confused and illiterate world, and it gained more focus and attention, which made it grow to dominance, fed and led by agriculture. However, Adam and Eve is just a story with no proof at all, a myth, a metaphor; in fact, in Hebrew the word Adam means mankind, and the word Eve means life (the same as Isis). So metaphorically, Mankind was tempted by Life to discover for itself what was good or evil, claiming free will and independence from their god-parents. Taking Adam and Eve as meaning Mankind and Life does not rule out our own biological evolution at all: "God created Mankind" - that's exactly right - by 3.4 billion years of evolution. It also makes more sense that Eve (Life) was created first and brought forth Adam (Mankind), but the story was written by patriarchs who just could not have it that way.

Taking the DC's metaphoric creation story further, Cain symbolizes the agricultural way of life taking over Abel's Paleolithic way of the nomadic foragers. Cain killed the old way of life, and agriculture usurped the hunter-gatherers, put the food supply under lock and key so there was no more free food, which forced everyone to work within the new system in order to have the money to eat - or die resisting.[4] The religions of WC, of agriculture, believed they knew the right way to live and have forced it onto everyone ever since, taking lands by any means possible, lands the DC deemed were being "wasted" by not being used to plant crops, as if development is the only and wisest use of land. The winners have written history ever since, creating a story that assumes human history only began six to ten thousand years ago and that pre-agriculturist people were miserable and irrelevant, but, in fact, they are the ones who remained in "the Garden" and had the best, most free and connected lives. No indigenous populations have ever willingly given up their ways of life to join the DC; they have never seen or thought the DC to be a better option.

Nevertheless, we are where we are, and mankind would not have evolved to its current level of technology without having been changed by the effects of agriculture. It's not about going back to "living in caves," and it's not about "destroying Western Civilization" as some have expressed in great detail.[5,6] That's simply and realistically not going to happen, not unless nature does it or we self-implode from some horrendous nuclear or economic disaster or other. It *is* about admitting our mistakes, learning from them, and changing what we've been doing wrong so we will heal ourselves from continuing to be a cancer upon the Earth and evolve our culture before it's too late - which is *now* – with a quantum leap of *conscious evolution, which we are now capable of.*

It's definitely time to update and evolve our understanding of God so we're not stuck in fundamental religious mindsets that originated thousands of years ago before we ever even understood the Earth was a sphere orbiting around a star. The Roman Catholic Church vehemently expressed that the invention of the printing press was "the work of the devil" because they saw the education of the masses as a huge threat to their institutionalized power, control, and wealth. It was the work of the Church that kept people ignorant and uneducated about anything other than what the Church wanted them to know, psychologically blackmailing people to keep people stuck in a mindset of religious beliefs that were set in stone, the stone of unchanging minds that are killing our Earth.

WOLVES

THE TERM TROPHIC CASCADE describes the domino effect that occurs in nature when as little as one link in the food chain is disrupted, which in turn causes the loss of other food sources and the resultant disappearance of several tiers of species. Trophic cascades happen when an apex predator species is extirpated from its native habitat. Wolf eradication is a classic example of an anthropogenic trophic cascade.

The last wolf was killed in Yellowstone National Park in 1926. After that, the elk herds flourished without fear and began to browse in groves and riparian corridors where circumstances had previously been too dangerous for them. The aspen forests were eventually wiped out because the elk ate all the aspen seedlings, so none were left to replace the dying elders. This ultimately left beavers without the means to perform their function, causing wetlands that supported a plethora of plants, amphibians, birds, and insects to be extirpated from the ecosystem.[1] Meso-predators, such as coyotes, bobcats, and foxes, which were controlled by the wolves, grew in population and wiped out smaller prey already diminished by forest and wetlands loss. Songbirds were lost, and even eagles moved on. Without riparian tree shade, native fish populations suffered from a rise in temperature of their waters. The entire landscape and food chain was devastated by the removal of wolves. Predators not only control prey; they also indirectly regulate biological processes down the food chain.[2] Without wolves there was a collapse of the ecosystem, from aspens and beavers to birds and frogs to butterflies and dragonflies to fish and cattails. Man had to step in to cull the elk herds, but it didn't work because elk continued to browse where they normally wouldn't if wolves were present. As soon as wolves were reintroduced in the early 1990s, the ecosystem began restoring itself back to a healthy balance.

However, the old ideology that embodies the hatred and fear of wolves persisted, and wolves were soon nearly sent into extinction by bounty hunters, ranchers, trappers, and government agencies even resorting to helicopter hunting until wolves were legally protected under the Endangered Species Act. A national effort to then reintroduce them back to their native habitats was successful enough to recently remove some wolf species from the endangered list, and immediately certain factions of the DIC got the helicopters out again, declared a renewed war on them, and started hunting, trapping, and poisoning them in vehement earnestness without regard for wolves' eternal extinction.

Wolf populations are naturally controlled by the amount of prey available, heaven forbid it be a cow on BLM land where a cow shouldn't even be, a cow whose death is reimbursed to the owner by the government. Plus the environment is ruined far more by cattle than even by the loss of wolves. More wolves, less cattle equals a healthy environment with more wildlife, but that's not something the DC can even fathom because the land is not being put to the DC's version of so-called good use, that is, it's not being fully exploited to make big money raising food for more people to live, necessitating more cattle…. To the DC, wildlife is an expensive, nonessential luxury for those "elite, city-dwelling, commie-lovin', candy-ass environmentalists who don't know squat and are interfering with our freedoms and lifestyle." That mindset is not going to voluntarily give up the deals and land control they have, even when it's killing the ecosystems they live in.

CATTLE

C ATTLE ORIGINATED IN THE Caucasus Mountains and were known as aurochs. These wild cattle were domesticated about 7,000 years ago in Mesopotamia (Iraq/Syria/Turkey). They were first used for holy sacrifices, and once they got to Europe, eventually cattle became a symbol of wealth. In 1493, livestock were brought from Europe to the Caribbean Islands, then into Mexico from Cuba in 1521 and into what is now the United States by Spaniards in 1540. As the DC spread into the Americas, formerly poor colonists could raise their own beef and finally eat like the upper classes of Europe.

When cattle came to the Americas over 400 years ago, they became an invasive species and have been changing ecosystems ever since. That's seventeen human generations after European settlement introduced their primary, domesticated food source, which has extirpated hundreds or more of species. Seventeen generations is a long time for things to become normalized and accepted; not many see anything wrong about the omnipresence of cattle outside of urbanized areas, but the Earth has continued to absorb and manifest the severe damage of their centuries-long alterations.

Habitat destruction is the number one cause that threatens plant species with extinction; livestock grazing, a specific form of habitat destruction, is the second cause. As cattle eat the native grasses and forbs, inedible weed species take over: the fourth-leading cause of any species becoming threatened with extinction is invasive species. Cattle cause all three of the top-four environmental damages and more.

Cattle-grazing destroys and changes entire ecosystems. Cattle cause practically every inch of ground to be trampled, turned to dust, and compacted, wiping out many species of all kinds. They either destroy river banks by flattening them out or create deep gullies from vastly increased erosion; either way, riverbank trees are lost. Cattle turn pure water (springs, creeks, rivers, lakes) into open sewers; make rivers more shallow and warmer, extirpating native fish; eradicate the native grasses, plants, and shrubs, allowing inedible weed species to invade and take over; decrease soil moisture and humidity, making microclimates more arid; win the competition for food over native ungulates, crowding them out; and cause apex predators to be exterminated to protect them, changing ecosystems drastically. The forested grasslands of Western landscapes are far from what they used to be. In short, cattle-grazing is causing massive desertification of millions of acres of land.

The Intermountain West (IW) of the U.S. is the area between the coastal mountains of the Pacific and east to the continental divide, stretching north-south from Canada to Mexico; it is comprised of eleven Western states. Cattle in the IW comprise less than 3 percent of beef production in the U.S. from 11 percent of U.S. producers, yet they take up 525 million acres of the IW "rangeland" (as if its definitive use was only for livestock), which equals nearly 80 percent of the eleven states and over 25 percent of the entire U.S. lower forty-eight states. Most of this land is public lands managed by the Bureau of Land Management (BLM) and the Forest Service; additional lands are private, state, county, city, and Indian reservations.[1] Cattle are allowed in National Forests and, worse, in many National Parks, wilderness areas, and wildlife refuges where they absolutely don't belong. Wildlife is even killed inside National Parks for livestock's higher priority to graze. Colorado allows grazing with no restriction in all designated wilderness areas even though logging, mining, road building, permanent structures, and motorized vehicles are all banned. And "in the Yellowstone ecosystem as a whole, the ratio of domestic livestock to all wild ungulates combined... is greater than 2:1."[2] Except for lands too cold in Alaska, cattle are ubiquitous in every state of the U.S. Estimates are "that 800 million acres... or 40 percent of U.S. land area, is devoted to raising cattle..."[3]; the number rises to *75 percent*, excluding Alaska, when the land used for growing crops for cattle is included. (To compare, all the cities and roads in the U.S. add up to less than 5 percent of U.S. land use.) Including Alaska, growing food for livestock and grazing them makes up 60 percent of the US land area.[4]

No large herds of grazers existed in the Intermountain West for over 10,000 years before cattle were introduced. Though there were subspecies of bison (now extinct) across the continent, the giant herds of bison we know of existed only on the vast area of the Great Plains between the Rockies and the Mississippi River and from northern Mexico through Texas to the north of Canada's provinces of Alberta and Saskatchewan.

However, cattle are not just a simple replacement of bison as is commonly thought. (Bison never had to be eradicated except as the means the DC used to finally subdue the most powerful tribes of natives; plus, the DC couldn't conceive of eating anyone else's "uncivilized, inferior" food.) Bison were superior for the Great Plains because they didn't overgraze anywhere, could tolerate drought and hard winters, and didn't loiter around water to ruin riparian ecosystems. Cattle, on the other hand, lay waste to every environment they're put in. Grass-fed or feed-lot cattle, it doesn't matter; their impact upon the environment and other species is disastrous and appalling. Public lands ranching has destroyed more native vegetation, more wildlife and wildlife habitat, more riparian areas, more natural water sources, has caused more soil erosion, and more invasion of non-native species than any other land use.[5] For environmental reasons alone, there is no good reason to eat beef.

Cattle are water-dependent and tend to linger close to their water supply. As an example of what has happened, back in 1870, before the cattle explosion in southern Arizona, the San Pedro River had an abundance of timber with plentiful grasses, and the riverbed was shallow and grassy, its banks full of luxuriant growth, but after thirty years of grazing, erosion had cut the river ten to forty feet below its banks, and its trees and underbrush were gone.[6] And it has never recovered. ("In 1870, the total number of cattle in the Arizona Territory was only 5,000... by 1891 the population of cattle [there] had grown to an estimated 1.5 million....")[7] There are very few exceptions to this in the West: the EPA and the U.S. General Accounting Office have both reported that livestock are the major source of riparian degradation on public lands in the West, that livestock has damaged *80 percent* of the streams and riparian ecosystems, and that riparian areas throughout the West are in the worst condition in American history.[8] "Thus have ranchers destroyed the most productive wildlife habitat in the rangeland West."[9]

It's not just the structure of rivers, springs, creeks, and lakes that are ruined; it's the very water itself. Cattle defecate at will with no discretion whatsoever as to where and can excrete up to seventy pounds of their runny feces and almost fifty pounds of urine per day. Once-pure watersheds are now thoroughly filled with protozoan parasites such as *Criptosporidium, Giardia,* and *Listeria* as a result, plus a countless number of coliform and streptococcus bacteria.[10] "[A]lmost one hundred microbial pathogens harmful to humans [can] be found in livestock waste."[11] The USDA estimates that animals in the US meat industry produce a total of 1.4 billion tons (*2.8 trillion pounds*) of body-waste, which is many times the amount produced by the entire human population of the United States each year, yet the cattle industry provides no treatment systems for it. The cost of treating water and the cost of disease outbreaks caused by the livestock industry are negative externalities not paid for by the cattle industry. "American livestock contribute five times more organic waste to water pollution than do people, and twice as much as does industry,"[12] not including steroids and hormones in the waterways.

If all this isn't reason enough to quit eating beef, an incredible amount of fresh water goes into producing it. Beef production demands an estimated 3,430 gallons of water just to produce *one six-ounce steak;* "a typical meat eater's diet requires 4,200 gallons of water daily… [while] a pure vegetarian's diet uses only 300 gallons."[13]

> In California, the major user of water is agriculture, [which] accounts for 83 percent of all water used [there]. …Growing feed for cattle… [on] irrigated pasture and hayfields consume more water than any other single crop in California…. Together, alfalfa and hay and pasturage account for approximately half of all water used in the state. …For example, nearly 1 million acres of irrigated pasture requires… as much [water] as an urban population of 23 million. Pasture… is the single largest water user in California [and] is an extremely low value crop. …The story is similar in other western states.

In Colorado, some 25 percent of all water consumed goes to alfalfa crops. In Montana, agriculture takes 97 percent of all water used in the state, and just about the only irrigated crop there is hay and pasture forage; more than 5 million acres in the state are irrigated hay meadows. In Nevada - the most arid state in the country – domestic water use amounted to 9.8 million gallons a day in 1993. By contrast, agriculture used 2.8 *billion* gallons of water per day… and the major crop is hay for cattle fodder… while… wetlands and wildlife refuges… often go bone-dry. Most agricultural water [in the West] grows low-value crops.[14]

Springs, creeks, rivers, lakes, and artificial ponds (called tanks) have been altered all across the country to accommodate the needs of cattle first. "The diversion of water for livestock use has caused the complete drying of stream beds…."[15] Many dams were built just to serve the needs of the cattle industry. "Ninety-five percent of the land that the Colorado River Storage Project was to irrigate would grow alfalfa and other grasses to feed livestock, 'about the most unprofitable use which could be made of irrigated land.'" *Congressional Record*, "Construction of the Colorado River Storage Project," April 18, 1955.[16]

The lost opportunity costs spent on raising cattle are mindboggling. For example, about 1.5 percent of California's land is used to grow vegetables (about 1.5 million acres), which grows about half of all the vegetables grown in the United States[17], while 5 million acres are used in California just to grow alfalfa. Yet with California's water shortage problems, water has recently been legislated to city dwellers and away from farms raising food for humans, but it was never even considered to cut water usage for cattle and their feed crops. Many food crops, such as miles of almond orchards, have already completely died out.

People scoff at the idea of livestock contributing enough methane to influence climate change, but methane is much more potent than CO_2. There are about 1.3 billion cattle in the world, and together with all the horses, sheep, goats, camels, and pigs, they belch out over 100 million tons of methane per year, an increase of over 450 percent since

1900. In 2006, the United Nations Food and Agriculture Organization said that livestock was responsible for more greenhouse-gas (GHG) emissions than transportation. In New Zealand, half of GHG emissions come from livestock alone. Grass-fed cattle also emit 50 percent more methane than grain-fed beef because grass is so hard to digest.[18]

People that quit or drastically cut down on beef consumption create much less demand for it, lessening the damage of such an environmentally poor choice of food production. Less demand would temporarily lower the price, and some producers would be forced out of the market. If the U.S. government would quit subsidizing them, it could be the ones in the IW who have already destroyed those ecosystems. This one-quarter of land of the U.S. needs to be healed and used for better economic and environmental purposes. Fresh water systems need to be restored, certain dams taken down, grasslands and forests rejuvenated, and especially an end put to the waste of so much precious water; it would also help improve the air and allow native species to return. "Even if livestock grazing were to end today, some areas of the West are so severely damaged that the time required for recovery may be on the order of centuries, or longer."[19] An ecological nightmare even yet to wake up from, livestock grazing in the arid West is obsolete, a relic in today's world that needs to be properly ended.

The economies of the IW are not jeopardized if grazing privileges were to expire on public lands, and the environment would definitely gain. Only a tiny sliver of jobs (0.07 percent) are directly tied to federal-lands grazing, making federal grazing rather marginal economically[20]; however, beef production in the IW could be impacted by half (1.5 percent of national production), a cut which would allow at least 300 million acres to begin to recover.

The number of cattle reached 254.4 million on western rangelands in 1990[21] but is much lower now partly because, after so much abuse, the land can no longer support them. In Georgia, one-half an acre is needed to raise one cow, but in the IW it takes 25-50 acres, and in southern Utah, it takes up to one square mile (640 acres).[22] "The official destruction committed in the sacred precincts of this massive range would be called vandalism if others had done it. The damage is so vast, so incredible, so awful that it has a permanent effect." - Former Supreme Court Justice William O. Douglas, *My Wilderness: East to Katahdin*. 1961.[23]

The BLM was not set up until 1946, primarily to manage the cattle business every bit as much as the Forest Service was set up to manage the timber business, and it had only been twelve years earlier, with the Taylor Grazing Act, that any laws were made regarding the management of rangelands in the U.S. A prior system – a cattle culture - had already been established with about 400 years of domestic grazing, which makes ranchers believe the rangelands are essentially theirs. The Grazing Act allowed ranchers to buy grazing permits for the use of publicly owned federal lands. The Supreme Court as recently as 2000 stated these permits do not convey property or grazing *rights*, only privileges in ten-year increments.[24] The fees for the grazing permit are only the minimum $1.35 per AUM, or $16.20 per year per AUM. (An AUM is an Animal Unit Month, the amount of forage required by a cow and her calf or an ewe and five lambs per month.) This is an extremely low fee to charge to allow cattle on hundreds of millions of acres of public lands.

About three-quarters of grazing permits are given to large-scale cattle ranches that are owned by billionaires or corporations such as Anheuser-Busch, Texaco, and Union Oil, and they are all unprofitable enterprises kept alive by government subsidies.[25] Taxpayers are paying wealthy ranchers to severely degrade public lands. "Despite the cowboy's image as a rugged, independent individual... [t]he western rancher is dependent on what is, in essence, a welfare program."[26] "Cattle ranching on the public lands of the American West is the most [inviolable] form of public welfare in the United States." - Edward Abbey.[27]

The rugged individualism characterized by the cowboy and the Western rancher are deeply embedded in North and South American culture to the point of being a big part of national identities. (Even though the cowboy originated in Mexico and was a lowly job done by slaves or their equivalent, beef consumption has long been considered a sign of superiority, both sexually and racially, and connected with masculinity; manhood is still associated with eating meat today[28]). Very large special-interest groups have formed which have infiltrated political leadership, influenced public opinion, and have thus maintained or created legislation that is strongly in their favor. Such political influence, with its ensuing agencies of unfair self-regulation plus many grandfather clauses, allow ranchers to further degrade lands with cattle where they would not even be allowed under today's environmental laws.

Public lands in the U.S. are primarily managed for the livestock industry, an industry led by about 30,000 very wealthy and politically powerful ranchers who believe the country has an obligation to maintain their way of life. These "livestock barons" are economically benefitted by a loud and vocal cowboy-loving demographic, based on cultural mythology, which enables them to grow fat at the public's (and Earth's) expense. Ranchers and their supporters profess to detest "big government" and see themselves as models of rugged independence, but they rival other industries in the amount of government handouts they receive.[29]

The problem is that ranching is so deeply entrenched in the DC. The industry, government, universities, politicians, and banks keep public-lands livestock grazing in place and free of reform because they each have a vested interest in maintaining the hegemony of the industry.[30] The expectation that ranchers will keep their grazing permits indefinitely has allowed ranchers and banks to treat grazing privileges as private property when they use the permits as collateral for loans; subsequently, the rancher waives all rights to the permit in advance in case there is ever a loan default. Ultimately, it's the Forest Service (or the BLM) that holds the grazing permit in escrow until the loan is paid back to the bank; the rancher with the loan only holds the privilege to graze.[31] Billions of dollars have been loaned to ranchers this way. The fractional banking oligarchy makes billions more off the ranchers' sizeable deposits and uses their considerable political clout to oppose reforms that threaten their *modus operandi*. Furthermore, banks make loans on permits largely on the basis of the number of cattle the rancher is authorized to graze, pressuring ranchers to maintain high numbers of cattle even in times of drought or degraded resource conditions so the permits won't lose value.[32] Meanwhile, the Forest Service is very compromised in its mandated position to protect and restore the land, water, and biota required by the Forest Management

Act, Endangered Species Act, and the EPA.[33] This system of easy money also allows ranching businesses to artificially survive under the guise of a free market they would not otherwise be able to stay in. Lastly, many universities receive their support from the cattle industry and would cease to exist if they did not support the banks, ranchers, and government through pushing education that maintains the status quo.[34] While all these human machinations are performed, our Earth has taken on all the damage of such dysfunction.

"World meat consumption rose from 44 million tons per year in 1950 to 217 million tons in 1999"[35] or about 500 percent. It leapt to 253 million tons by 2003, rising in China by another 500 percent from 1977.[36] About sixteen pounds of grain and over 9,000 gallons of water are needed to make one pound of beef, while a pound of wheat requires only 60-130 gallons of water. "The caloric energy provided by beef is only one-seventh of the energy of the grain fed to them."[37] While 60 million people starve to death each year for lack of grain, more than 750 million tons of it is fed to livestock. "More than 70 percent of the grain grown in the United States and close to 40 percent of the grain grown worldwide is fed to animals...."[38] To quit eating beef would vastly improve human lives – even of the ones eating it.[39] Furthermore, the same amount of money we throw away on subsidies to the livestock industry could buy up, protect, or restore wildlife habitat and preserve endangered species.[40]

Privately owned livestock grazing on public lands have priority over wildlife, including endangered species. The U.S. Wildlife Services (part of the USDA) spends millions of dollars to kill millions of animals annually from prairie dogs to grizzly bears, and dozens of species in between, all to protect cattle, a service given to ranchers for free at taxpayers' cost. This is in spite of the fact that most predators never attack livestock. Predators only cause 2.7 percent of deaths to livestock. The other causes are respiratory problems (27.5 percent), digestive problems (19.7 percent), unknown causes (15.2 percent), birthing (14.8 percent), weather (9.5 percent), and other (9.1 percent). Only poisoning (1.1 percent) and theft (0.4 percent) are smaller problems than predation.[41] It would be more economical to abolish the policy of killing off wildlife and reimburse ranchers for any losses by predation.

As wildlife and nature then came more into balance, ranching on public lands could be phased out as it should be, given its long-proven track record of ecological disaster.

The environmental problems that come with cattle ranching are global in scope. In Brazil, for example, of all the Amazon rainforest that has been cleared, cattle inhabit over 85 percent of it; there, too, beef production is not profitable without government subsidies. The cattle population there has grown from 26 million in 1990 to 65 million in 2005.[42] Brazil is also the number one beef exporter of the world, and the market has grown by 600 percent since 1999.[43] As in the U.S., the land is being destroyed, the very prevalent cowboy mentality says, "it's my land, I own it, don't tell me what to do," (the formerly lush rainforest with thousands of gorgeous, endemic species is gone in the state of Mato Grosso, now 83 percent privately owned and fully devoted to cattle-ranching[44], surrounding a tiny reserve for native species). Also, vast agricultural resources are used to feed the cattle. Brazil is second in the world at growing soy beans (GMO and on former rainforest land), which go largely to feeding cattle. Clearing forests for cattle versus clearing for farming in Brazil has a 5.5 to 1 ratio.[45]

The same is true all over the world. "Nearly 70 percent of the deforested land in Panama and Costa Rica [a hotspot region] is now cattle pasture."[46] In Australia, converting coastal lands to pasture is contributing greatly to killing coral in the Great Barrier Reef with run-off silt.[47] "One half of the planet's total land area is grazed by cattle and other domestic grazing animals, and at least one-quarter of all cropland is now devoted to raising feed for livestock."[48]

"Hooved locusts," is what the lover of Yosemite, John Muir, called cattle, and he was right. Edward Abbey, an outspoken environmental author, called them "ugly, clumsy, stupid, bawling, stinking, fly-covered, [feces]-smeared, disease-spreading brutes. They are a pest and a plague. They pollute our streams and rivers. They infest our canyons, valleys, meadows, and forests."[49] And he was right. But, the cattle are innocent: it is man which is at fault for breeding and allowing cattle to be the way they are, it is humankind that has caused cattle to become ubiquitous upon the land, and it is mankind that has perpetrated severe abuse upon them. Abbey also said, "The rancher (with a few honorable exceptions) is a man who strings barbed wire all over the range; drills wells and bulldozes stockponds; drives off elk and antelope and bighorn sheep; poisons coyotes and prairie dogs; shoots eagles, bears, and cougars on site; supplants the native grasses with tumbleweed, snakeweed, povertyweed, cow[dung], ant hills, mud, dust, and flies. And then leans back and grins at the TV cameras and talks about how much he loves the American West."[50]

Truly, cattle do ruin everything in sight, and intelligence was bred out of them to make them easier to handle. When one realizes the vast amount of damage humans do to the environment with their cattle, beef is still an expensive food for the elite; it is made affordable to the multitudes of wealthy nations by economic wizardry, but the hidden costs of negative externalities - the costs to the health of our planet - are enormous.

PRAIRIE DOGS

PRAIRIE DOGS ONCE NUMBERED up to five billion and inhabited over 150 million acres in the U.S. from Texas to Montana and east through the Great Plains to the Mississippi River. Thought of merely as "prairie rats" (although they're herbivorous members of the squirrel family), it's still believed they can harm cattle. Coupled with their ability to disrupt farming, the prairie dog population, comprised of five different species, has been reduced by over 99 percent since the 1850s; two of the species have been on the endangered species list.

Prairie dogs are a keystone species, and the loss of them has changed entire ecosystems and the lives of many other species, species such as hawks, bobcats, foxes, coyotes, black-footed ferrets, owls, and many more that have lost prairie dogs as a food source, creating a disastrous trophic cascade for wildlife. From 1900 to 1998, 103 million acres of prairie dog habitat had been reduced to 750,000 acres, a decline of 99.3 percent. This reduction was especially severe in the late 1990s.[1] In spite of this, the USDA's Wildlife Services [formerly the Predatory Animal and Rodent Control Department] spent over $10 million to kill over 2.7 million native "pest" animals in fiscal year 2014: "322 gray wolves, 61,702 coyotes, 580 black bears, 305 mountain lions, 796 bobcats, 454 river otters, 2,930 foxes… 22,496 beavers… [and] 15,698 black-tailed prairie dogs… [and] kills many more animals than it reports."[2]

Besides having been an enormous food source for many species, prairies dogs' largest contribution has been its role in the water cycle. Prairie dog tunnels permeated millions of acres of arid grasslands and plunged to a depth of twelve feet in fertile topsoil, turning rich loam to the surface. The tunnels channeled water directly into the water table instead of only soaking the root zone of the grass, which in turn kept streams and rivers flowing. Without prairie dogs, the grasslands became continuously drier to the point of affecting the microclimate of the prairies and beyond.

The southwest has been especially affected. An all-out war on prairie dogs entirely extirpated them from Arizona. "Their removal is believed to be responsible for the conversion of six million acres of grassland to desert scrub in Arizona and New Mexico."[3] This was done mostly for cattle, which combined to dry up rivers, streams, and springs with their overgrazing and trampling; invasive species also then crowded around what waters were left, most notably with the nearly dry Rio Grande River in Texas. Woody shrubs and trees such as chaparral, mesquite, and juniper took over the grasslands, and what forests still remained from heavy logging were weakened by the ensuing drought in an already thirsty land. This led to an explosion of bark beetles, killing the trees and leading to larger, more intense forest fires, which in turn created even more drought, as water tables continued to drop, drastically increasing desertification. All this happened because some people consider prairie dogs as vermin, not realizing they are essential to a healthy environment, yet we now know they are critical to biological and ecological wholeness. However, the minds in charge hold steadfast, as corporate-serving politics differ from science; the prairie dog is still considered an agricultural pest requiring mandatory elimination in several states[4], and increased desertification, loss of wildlife, loss of water, invasion of non-native species, and death of forests continues.

RELIGION, PART II

"The Latin church is the great fact which dominates
the history of modern civilization."[1]

RELIGION HAS HAD A fundamental part in shaping the ideology of
the DC for thousands of years; it has had a continuous part in leading
the world into our ongoing mass extinction, an extinction of species that
ultimately threatens each of our lives. For this reason, it is essential we
understand what got us where we are, so we can change our mindset
and behavior and let go of our destructive paradigm.

There is no better example of the destructiveness of religion than
examining the history of the Roman Catholic Church (RCC); it formed
the world's ideology more than any other institution by dominating
Europe for centuries. The ensuing domineering influence of Europe
ultimately spread throughout the world with its authoritative dogmatism.
There is plenty of evidence that makes the history of the extremely
suppressive Roman Catholic Church very clear.

It's not the individual laity of Catholics at fault, except for their
blind support of the Church, usually due to being born into a family
of the Church culture and causing a corresponding normalcy bias; it
is the historical policies of the institution of the Church that have been
intensely inhumane and criminal, and it has had a profound effect upon
why things are the way they are today. This does not absolve other
religions of the world; in fact, the worst factions of Islam are picking
up where the RCC left off centuries ago, both Abrahamic religions.
However, European theology is the one that took over as being the
dominant ideology: Western Civilization.

If it was not for the RCC, our planet would not be seen as a thing, like a machine full of commodities, given to humans to do with as they wish, which has caused abuse so severe the extinction of half of non-human species is probable in thirty to eighty years. Women would not have been seen as causing "original sin" and bringing evil to the world, treated as actual slaves for centuries, regarded as sub-humans, and subservient to men. Humans would not have believed having as many children as conceivable to be a religious duty, or that humans are the most important thing in the Universe, so anything and everything imaginable must be done to create and provide for as many humans as possible forever. Freedom on all levels and scientific and cultural advancements would not have been thwarted at all costs for centuries. The holocaust of burning nine million innocent women (and men, children, infants, and pets) at the stake after being tortured would not have happened; certain religious wars would not have taken place; continents would not have been robbed of all their primary wealth, with millions of slaves being used to do so and dying short, tortured deaths; and cultural trauma from fifteen centuries or more of such terror would not have become part of the memes of the human population, all sanctioned by the Roman Catholic Church. All of these events and more still have vestiges in our world today; indeed, they created the paradigm we live in - and we believe it's acceptable, normal, and unchangeable because it has persisted for so long, but only by the coercive mind control and terror brought on by a very obstinate Church.

It was not until 325 AD, with the Council of Nicaea, that the authoritative establishment took place as to which books (of many) were to be included in the bible. (The definition of bible is simply a collection of books). Constantine, the first Roman emperor to become a Christian, presided, and historical witnesses wrote that he ultimately made an arbitrary decision as to which books were "divinely inspired" and which weren't; additionally, the bible was written with an acknowledged 150,000 mistakes.[2] Nonetheless, a large amount of bishops gathered from all across the Roman Empire at Nicaea to decide what the official uniform story about Christ was to be told. The Council of Nicaea – and there were many Councils over the years that regulated the story and policies of the RCC - politically determined that Jesus was the divine son of God over three hundred years after his alleged life and death. From that point on, the RCC continued to grow, becoming recognized as a legal entity, with the Edict of Milan in 400 AD, and as the religion of the state. Ten years later Rome was sacked and the fall of the Roman Empire began. Europe descended into chaos, for there was no longer any central form of government. The Roman Catholic Church filled the void: the Roman Empire was replaced by Christendom, led by the RCC, which took Europe swiftly and indisputably into the Dark Ages.

The Dark Ages, absorbed into the name of the Middle Ages to comply with modern "political correctness," happened when Europe's cultural roots were forgotten and lost by Europe's separation from the cultures of its origins, the Grecian and Roman civilizations. Even the infrastructure of roads, farms, and aqueducts were left to deteriorate, and the arts were abandoned. The Dark Ages lasted a very long one thousand years and further traumatized the people of the continent worse than any other time in Europe's history. For five hundred years, "from the seventh to the eleventh century no individual thought can be traced"[3] due to the RCC's disallowance of it, enforced by means of torture and death. Panicky religious superstition ruled the times, and science or any progressive thought was judged by the Church as something evil and subversive. Christianity was barbarically forced upon people everywhere by the RCC. Many in today's world *still* believe that science is "the work of the devil," such as the genetic proof that man evolved from other primates and that our Earth is older than six thousand years.

As the Middle Ages continued, the Crusades and the Inquisition arose, adding nine more centuries to the six previous centuries where freedom was explicitly quashed. The Crusades were church-sponsored wars against Muslims in the Middle East that officially lasted from 1095 AD to 1270 AD (almost two hundred years), although more Crusades continued into the 1400s.

Pope Gregory IX instituted the Inquisition in 1231 AD, torture was first authorized by Pope Innocent IV (quite the oxymoron) in 1252, Pope Sixtus IV began the Spanish Inquisition in 1478, the Inquisition spread to Sicily in 1517, Pope Paul III started the Roman Inquisition in 1542 to combat Protestantism, and the last of the Inquisition did not end until as late as 1834 in Spain.[4] The RCC used 600 documented means of torture for 600 years of the Inquisition after the first 600 years of the Dark Ages, with the Crusades in between. (Overlapping the Inquisition, the Renaissance slowly re-introduced Europe's higher cultural roots.) For at least sixty generations, during a ruthlessly brutal 1500 years, people lived in fear for their lives. To save their own lives, neighbors and even family members reported each other to Church officials and shunned anyone who was suspected of wrongdoing by the Church so as to avoid guilt by association. Suspect women who had nowhere to go were left alone to fend for themselves out in the wilderness.[5] Treachery ruled the day, ruled the centuries. It became a continent of "every man for himself," leading to the full loss of a bonded, trustable, collectivist society and the beginning of a very fractured, opportunistic, individualist society.

"Christianity is the great enemy of the natural world...."[6] The Christian faiths have forever propagated the belief that man is the culmination of creation and that the world was given to man to do with as he pleases, to have dominion over the Earth and multiply our species, and that the Earth was here solely for the sake of man. ("God put those trees on the hillside for people to use, and to just let them rot or burn up is going against His teachings..." is the dominant worldview of most natural resource managers, loggers, ranchers, miners, and other commercial users of the land[7]). The Earth and its biota became a thing, a commodity to be used to increase our numbers. Women were here to help man and propagate the species, no more. This ideology has become extremely problematic, as human overpopulation is the number-one reason for every current problem concerning the environment and the human-caused ongoing mass extinction; habitat destruction is the number one reason for species going extinct, as man uncaringly exploits the Earth for his own gain as much as possible. "We have no divine right to treat all other life as 'resources' for our use."[8] We only convinced *ourselves* of that "divine right" out of greed, convenience, and hubris, realizing nothing was stopping us.

"The true enemy of freedom and progress has been the church."[9] The RCC has fought hard to keep the populace in the dark; an uneducated laity is easier to control. The Church has always opposed all new discoveries of science and all reforms of any sort. When the printing press was invented, the RCC vehemently declared it was the work of the devil. For centuries, little was accomplished in medicine, as the pursuit of scientific truth was seen as a threat "and was thus severely constrained by... the Church."[10] "The careful student of history will discover that Christianity has been of very little value in advancing civilization, but has done a great deal toward retarding it."[11] The RCC didn't recognize Galileo's theories until 1800, two centuries after it had shown him the torture chambers, threatening him to keep silent about his discoveries. "Christianity has ever been a religion of the emotions rather than of the reason. The former was cultivated; the latter bitterly condemned. The church has ever found its most powerful enemy in reason, hence the exercise of reason has ever been a crime in her eyes."[12] The contemporary position of the RCC over birth control is essentially no different and is creating enormous problems.

The incessant indoctrination of the Church, preaching for centuries that Eve introduced sin and death and all evil into the world and got man kicked out of paradise, created and perpetuated extraordinary abuse upon women for almost two millennia. Every other culture ever known has treated women better, including ancient Rome and Paganism. The RCC taught that women had a special curse put upon them by God, deserving of the punishment given out by men; they thought Eve was Satan's daughter as much as Jesus was God's son. Married women were subject to beatings with no access to outside authority. Women in England were for more than a thousand years legislated for as slaves, and wives were bought and daughters sold. Under the iron-fisted rule of the RCC, women were merely tolerated as a necessary evil for the purpose of bearing children, and women were disallowed the right of inheritance nor any political or household authority.

> Had not man been trained by his religion into a belief
> that woman was created for him, had not the church
> for eighteen hundred and more years preached woman's
> moral debasement, the long course of legislation for
> them as slaves would never have taken place, nor the
> obstacles in way of change been so numerous and so

persistent. ...Unless an heiress, woman possessed no social importance; unless an inmate of a religious house, no religious position.[13]

In all the time since the inception of the Roman Catholic Church, women did not receive the right to vote - even in the freest country in the world at the time - until 1920, just ninety-six years ago in the U.S. Even then, women still had to fight for equal rights in the 1960s and on, all the vestige of patriarchal Church indoctrination. How this relates to the HOME is that women only hold a small minority of positions of power, and even then they have to obey and work within the patriarchal system, if they're not actually converted patriarchs themselves; feminine energy would not treat our Earth the same as the ruling class of men do.

Many controls over women were put into place under Christian law regarding marriage. A custom known as the marquette, which was upheld by the Church and the state for hundreds of years in all Christian countries in which feudalism existed - including Scotland, England, Germany, and France - allowed the feudal lord to take any freshly married woman to his bed for up to three nights and was entitled to any first-born on the assumption that the child was his.[14] This policy of rape and humiliation did not begin to be abolished until the late 1500s.

Marriage in the Middle Ages was looked upon as a necessary evil, so in a custom that lasts to this day, the church only recognized marriage if it was sanctioned by a priest; otherwise, the married woman was declared a concubine with the potential of Inquisitional consequences. Thus the church had control of the family (as it still does to lesser degrees). Once married, a woman and her real and personal estate and "her children became her husband's property and part of another family; she was entirely lost to the family of her birth." The wife was the husband's personal slave, all her time and the fruits of her labor were absolutely his.[15]

Included in that time period was the burning alive of up to 9 million people, mostly innocent women convicted of "sorcery" and "heresy," which meant anything that wasn't agreeable to the Church. "That a woman should be burned alive for a crime whose only punishment for a man was a few months imprisonment, was in unison with the whole teaching of the Christian Church regarding women."[16] Considering the percentage of the population killed, this is a holocaust greater than that perpetrated by Nazi Germany, and though it is taboo to speak of Nazis, the RCC still flourishes tremendously in spite of this and all the rest it has done.

In the eighth century, the Christian Emperor Charlemagne was the first to approve of torture for accusations of witchcraft. "Torture was rapidly adopted over Europe, and soon became general [policy] in the church; the council of Salsburg, 799 [AD], publicly ordering its use in witch trials." Children and pregnant women were not exempt from torture.[17] Sometimes children were tortured in front of their mother in order to get a "confession" out of her. The Church grew enormously rich, as it confiscated the property of the condemned, sometimes splitting 50 percent of the riches of the estates with judges and prosecutors. "The love of power, and the love of money formed a most hideous combination for evil in the Church…."[18]

The word "witch" is from wekken, to be intuitive, prophetic, psychic, have healing powers, and (to the Church) to have a strong will. "In the fourteenth century the church decreed that any woman who healed others without having duly studied, was a witch and should suffer death…. Few women dared to be wise, after thousands… had gone [to their torturous deaths]."[19] So traumatic was this time that a man and/or his sexual anatomy being called a prick in today's world, 400-1200 years later, refers back to the "Witch Prickers": men who stripped all clothing off any age of female, shaved all hair off, and "minutely examined all parts of her body for the devil's sign," which could be something as small as a mole.[20]

"The world has produced no system so thoroughly calculated to extend its own power and wealth… which, under the guise of religion, appealed to man's superstition, and ruled his will under the assumption of divine authority…."[21] "The preacher… occupies, according to him, the next place below the angels."[22] To become a priest, a man had to be able to read and write. At the time, only the "nobility" were literate. Since the priesthood claimed direct inspiration from God and taught they were thus infallible, it drew all kinds of unscrupulous men to the clergy who realized they could do anything for personal gain, malice, and lust and get away with it. Religion became the primary system of obtaining power, and the church pandered to those with money and power for its own aggrandizement, stopping at nothing to secure it, setting the entire continent of Europe in a state of moral corruption.

> The claim of infallibility... became all-potent when advanced... under the guise of religion, into which the element of fear [was] largely entered. To disobey a priest was to endanger [one's life and] salvation; it was libelous and treasonable to question the purity of a priest's motives, hence religion became a screen for all vice and a source of moral degradation to all women. In the seventeenth century it was a proverbial expression to say, "As corrupt as a priest." [Popes and bishops were not immune]: Pope Pius IX had two [daughters] from two of his mistresses; Pope Gregory XVI [had many] mistresses, one of whom was the wife of his barber...[and] he was one of the greatest drunkards of Italy. Pope John XIII...was proved...of having been guilty of fornication, adultery, incest, sodomy, theft and murder... [and] he had seduced and violated 300 nuns. Henry III, bishop of Leige... [had] sixty-five illegitimate children.[23]

Replace pope, priest, and religion with legislators, lobbyists, and certain corporate officers, and you have the modern version of the same basic dogma reoccurring today. The same patriarchal characteristics still exist: love of power, greed for money, total self-interest, and disregard for the rights of others. In addition, canon law became thoroughly incorporated into English common law, which in turn became the basic law of the United States. The Church and the state have been in an inseparable collusion for centuries. Theological and political ideology merged as one. "Ideologically, it was impossible that the churches could judge the actions of the government or that the government would become hostile to the churches. ...If one fell, the other was sure to go. ...With American Christianity and the American government largely welded together in a holy alliance, a threat against one would appear to be a threat against the other."[24] This is still true to this day.

The Council of 1215 established the use of oral confession. This allowed the priesthood to gain further access to "all family, social and political secrets, thus acquiring information whose power for evil was unlimited." Additionally, 100,000 women in England alone were seduced by the priesthood through confession.[25]

It's no wonder that Protestantism - religious protest – came about. The Catholic Church could by no means live up to any of its own Christian principles or Ten Commandments. That only the pope could hear God and knew what God wanted the people to hear was total hypocrisy, as it was well-known the popes were the ones ordering the worst of human cruelty to be carried out in the name and love of God - and growing very rich while doing so. Catholic and Protestant countries have been fighting each other all over Europe ever since, the latest having been recently between Ireland and England. European countries, cultures, families, and individuals were all severely fractured in all ways, and they brought their perfidious culture to the rest of the world as surely as they tried to escape it.

The pope, by his own authority, actually gave the continent of Africa to the Portuguese crown and the continent of most of South America and most of southern North America to the Spanish crown, saying it was because any land without a settled form of government is open to claims, and that "the natives are not humans."[26]

By the time European refugees started escaping to the Americas, Europe had lived an absolute nightmare lasting over a thousand years under the rule of the RCC and had long since lost all its innocence, but they brought their understandably psychotic mindset with them. Eight of the original American thirteen colonies recognized witchcraft as a capital crime. In Europe, "(a)ll humanitarian feeling was lost. Mercy, tenderness, compassion were all obliterated. Truthfulness escaped from the Christian world; fear, sorrow, and cruelty reigned preeminent.... Freedom was an unknown word... under the ban of the church and the state."[27] This is what came over the Atlantic in increasing numbers, seeking healing from the heart of the world.

Our world desperately needs people to evolve their religions. The RCC, for example, is still defending its precepts from 1,700 years ago and still defying science. It's not the only one. Mormonism, for instance, the latest of the Judeo-to-Christian-to-Catholic-to-Protestant progression (just like evolution through speciation by the way), has done nothing to help: it still believes that evolution is false; still debates whether the Earth is seven thousand years old or not; that the second coming of Christ is due; that there's another planet for them to live on, so it doesn't matter whatsoever what happens to this one; and to have as many babies as they can like Brigham Young, who was the first Mormon president and had nineteen wives and forty-two children. If it wasn't so otherwise provable, the RCC would still insist the world is flat and burn the famous astronomer, Bruno, at the stake again for saying the stars are other suns like our own with planets.

Humanity suffered a horrendous childhood and adolescence, and the abusive patriarchs everywhere who brought it on deny their part in it and refuse to change, continuing their behavior only in more modern ways, but still keeping governments, humanity, and personal will in tight control. The RCC and others bear an enormous responsibility for our ecological crisis, which they even continue to deny or defer all free will and responsibility to being "God's plan" when convenient. It is *highly* immoral behavior, especially to cause, but also to allow, God's creation of life on Earth to be destroyed in a careless and greedy, human-caused mass extinction of species, adamantly and blindly holding fast to beliefs that were formed by religion *"asserting as true things of which it knew nothing"*[28] thousands of years ago to present day.

OBSOLETE

D ENIAL GOES ALONG WITH the fairy-tale thinking that the way we're doing things can go on forever, that the way we are living is the right way, that tradition must be held on to at all costs, and that if some of those costs are the mass extinction of species, so be it, because human progress is all that matters and is essential to our survival. Denial, to the point of mental illness (since mental illness is defined as being out of touch with reality), doesn't realize what is really going on and must be cured.

In the past, we could get away with environmental denial because there were always new territories to exploit in the name of progress, but now the world is full, and we are over-extracting our natural resources by up to 50 percent annually. Rivers are running dry, aquifers are lowering without sufficient replenishment, deserts are expanding and being newly created, plastic (oil) and radiation is filling the ocean, and so on. Progress has reached its apex and is now giving back obvious diminishing returns environmentally. Yes, the way we are operating can continue, but it will have drastic and severe results. No one wants to believe this because the warnings have been heard before and still nothing has happened quite as predicted. Plus there's money to be made; the economy is all that matters. Fundamental ideologies still believe we are separate from nature or can overcome it when we must; we'll always figure something out, especially with AI dawning.

We don't recognize or admit that many ways of doing things are obsolete. For instance, the most contentious issue that causes the most emphatic debate and denial: having big families with lots of kids is very obsolete. Nor is it about misanthropy, which is also obsolete and doesn't work. We've got a real problem with overpopulation, so how do we solve it when laisse faire or voluntary policies don't work, and no one agrees on any proposed forms of control on the issue? And yet, the problem *must* be solved. It's the primary cause of our human-caused ongoing mass extinction; a misguided, obsolete ideology has already propelled us too far, and there are no good answers. Does it have to come down to one child per couple and one child per person for life, huge tax incentives, or forced sterilization? The whole subject is taboo. By what acceptable means is our population going to be brought to a level that's conducive to the health of our planet for the long-term? Just keep having innumerable babies until nature takes its course? The results of that policy are not acceptable either. Irresponsibility on this issue rules the world and is causing the HOME. We'd rather cause a planetary *mass extinction* than take *any* responsibility for our part in it. Instead, it's God's fault, nature's fault, the government's fault, religion's fault, and yet, hell beyond measure looms before us. Did God *really* say and mean to be so fruitful and multiply so as to obliterate His Creation of Paradise on Earth? Why don't people pay more attention to other parts of the bible which say not to damage the Earth? We have been very irresponsible with our natural procreative freedom, and the time of that irresponsibility must end. Our human overpopulation is killing species all over our planet. It's more obsolete to be against birth control than it is irresponsible to have an abortion. A self-preservationist, intelligent species would admit this and figure it out. Otherwise, nature will take care of the problem for us, but it will cost the horrendous extinction of millions of species and the increased suffering of even more billions of people.

Exploitation is obsolete. Though it is still happening, exploitation has peaked; the "principle" of the world's "bank account" of resources is being withdrawn at ever-increasing rates. It won't work. Payment of the environmental debt is beginning to manifest. Industry is especially wasteful and uses 233 percent more resources – water, oil, coal, etc. – than all the humans in the world (a 70 to 30 percent ratio).[1] This means that all the people in the world could recycle 100 percent of their recyclables, and it wouldn't even cut industry's waste in half. The businesses of industry are in too much economic competition with one another; they can't afford to clean up their operations. There are too many MNEs without loyalty to any place on Earth, so it's easy to exploit, to grab and go and slash and burn as fast as they can in order to meet the demands of the human population and make killer profits (literally).

Living irresponsibly during the human-caused mass extinction of species is obsolete.

Cattle ranching on 535 million acres of public land in the American West to produce 3 percent of American beef is obsolete. The land has already been destroyed and is becoming ever more desertified and full of inedible weed species. Ranchers these days are not the family style ranchers we're led to believe run everything; they are a very small percentage compared to the giant corporate "ranchers." Government subsidies keep the industry alive. The land can't even support the animals on its own; forty percent of precious farmland and water is used to grow crops like corn and hay just to supplement what little the cattle can graze on in the open range and in feedlots. "Livestock grazing in the arid West is as outmoded as is whaling in today's oceans." - Doug Tompkins.[2]

Doing nothing to help heal our Earth is obsolete.

Believing in religions that say the world is only 6,000 years old, etc., is obsolete. Six thousand light years of time and distance doesn't even come close to reaching the center of our own galaxy.[3] This is as measurable as the Earth not being flat. "Oh. Well God only made the Earth *look* round...." People could upgrade their religion without conflict by learning from a brilliant teacher in pastor David Dowd.[4,5]

Believing that humans are created separate from the rest of life on Earth never even made sense and is obsolete, as is that the Earth was given to us to do with as we please.

Believing that economic growth is the most important thing on Earth and overrides keeping the systems of the natural world functional is obsolete. There is no such thing as sustainable development in today's economic world. It is a propagandist lie that industry has told so they can continue with business as usual, which is destroying the functions of our Earth.

Multiple use is another fallacy that has been told in order for the Forest Service to maximize economic benefit; they can sell the timber, lease the land to ranching and mining companies, sell permits to tourist companies and hunters and trappers, and charge people for a plethora of mostly damaging recreational use (esp. with the sheer amount of users). Where do the non-human species go?

Driving personal vehicles fueled only by gasoline or diesel is obsolete. We could easily have electric or biodiesel or other vehicles if we demanded them. The technology has been with us for a long time, but auto and oil industries will not relent because not enough people are demanding loud enough, and the higher politicians are the industry's bootlickers.

Countries spending trillions on war and weaponry per year is insane and obsolete. The world could be healed and species kept from going extinct with even half that much money.

Business as usual is obsolete.

Using fresh, drinkable water to flush toilets is obsolete.

Golf courses are obsolete. What a waste of water and land.

Clearcutting forests is obsolete.

Big-game hunting, eating Bluefin tuna, using bear bladders, tiger bones, and rhino horn for questionable medicinal purposes, and using elephant ivory for anything is obsolete.

Driving big cars and being an over-consumer of anything just to think you impress anyone or to feed your ego to think you're superior to others is obsolete. All that extra money you blow to so-called look good could go to something much more worthy. To those of us who know it's obsolete, you just look like ignorant people who are mindlessly aiding and abetting the abusive DIC and the poison and propaganda it spews out.

Being an avid member of the DCI is obsolete.

Building large homes is obsolete. It's no longer a sign of success; it is a sign of exploitation and mental illness. How many acres of forest go into building those monstrosities just so people can show off or be way too comfortable at the expense of all other life? Development at all is obsolete; rehabilitating what we've already built is all our Earth can handle.

Our definitions of success are very obsolete.

Using nuclear fission to produce electricity is obsolete. You don't even know the truth about Chernobyl and Fukushima or the hundreds of other nuclear accidents never reported.

Suppressing "free energy" technologies, such as many of Tesla's patents, is obsolete. All the world's energy needs could be supplied by these clean inventions.

Homo sapiens need to *grow up.* The entire DC paradigm is obsolete. As long as we believe in and keep trying to function out of what is obsolete, which equals lethal ignorance, apathy, and dysfunction, it will send our Earth deeper into the anthropogenic mass extinction of species and our own demise.

RELIGION, PART III

"Such matters [of the Vatican] cannot be treated as matters of no substance or importance. They have impacted all aspects of life – moral, spiritual, political, and economic – at the turn of the twenty-first century"[1]

T HE WRONGS AND WOES of the RCC are not all in the past. It's mentionable because the RCC has been dominant in the world for at least 1600 years. Although the followers of the RCC are outnumbered now by almost half a billion Islamic followers (1.6 billion), and they both maintain and embody the errant DI that's causing the human-caused mass extinction of species, the RCC has shaped our modern world more than any other religion. The RCC, centered in the Vatican, is "an institution with 1.2 billion followers, 6 million lay employees, 4,500 bishops, 412,000 priests, and 865,000 members of religious institutes and schools."[2] This means about one out of six people in the world are Catholics, sixty million (one out of five) of which live in the United States. Along with the wealth it has amassed, the RCC's political influence upon the entire world is immense. Despite all the good people of the laity the RCC has turned out, the institution has an extremely abhorrent history even in modern times. Despite its help to millions of the poor, the Church has also caused most of the poverty through sanctioning past colonialist resource extraction and by its strict and global anti-birth control policy, which has substantially contributed to overpopulating our world, the number-one cause of the HOME. On the other hand, the Vatican has also caused tens of millions of human deaths in its support of war and innumerable crimes, deaths, and the great undermining of society from its enabling of the Mafia's financial empire and narcotics trade (not to mention its political influence).

Italy rebelled against the RCC in 1870 and retook territories within Italy the Church had previously stolen. The 108.7-acre Vatican was thus established within the city of Rome on and around Vatican Hill and was designated as an international country legally independent of all others, including Italy, with rules of its own and answerable to no outside authorities.

"With the loss of land [in 1870] came the loss of taxation."[3] With the loss of tax revenue, by 1929 the Vatican was broke and severely rundown, so on February 11, 1929, the Vatican signed the Lateran Pacts, a deal with Benito Mussolini that caused the rise to power of the WWII Italian tyrant, Hitler's ally. In return, the Vatican received, in all, the equivalent of $1.3 billion in 2014 dollars.[4] An undisclosed large sum of cash also went personally to the pope.[5] The Church's "ecclesiastical corporations" were granted a tax exemption.[6] Catholicism was declared the official and only religion of Italy. The RCC made this agreement with Mussolini even though he "...described priests 'as black microbes who are as deadly to mankind as tuberculosis germs.'"[7]

The Vatican signed a similar treaty with Hitler on July 1, 1933, titled the Reichskonkordat, which included a 9 percent tax from the paychecks of all working Catholics in Germany, and the $100 million per year tax proceeds (at 1930s value) were sent directly to the Vatican.[8] This deal with the Vatican put Hitler in power, who was raised Catholic, attended a monastery, and had at one time wanted to be a priest.[9] The RCC persuaded all Catholics to vote for Hitler as the Church had done for Mussolini. (The Vatican also supported another tyrant of the time, Spain's Francisco Franco.) Hitler's deal stipulated that "Catholics were now permitted to become full-fledged Nazis...,"[10] and it "required German bishops and cardinals to swear an oath of loyalty to the Third Reich."[11]

Not wanting to jeopardize its financial deals with Mussolini and Hitler, when millions of "undesirables" were being inhumanly exterminated during WWII, the Vatican never issued a single word of condemnation about the Holocaust while it was happening and for a long time after. "'Undesirables' were defined simply as all individuals who were not 'Aryan' and not members of the Roman Catholic Church."[12] Some Catholic priests even aided and supervised some of the mass executions.[13] It could be argued that WWII was a religious war perpetrated by Catholics mainly upon Jews.

At the war's end, the Vatican was the primary institution that enabled about 30,000 Nazi war criminals to escape via Church-created "ratlines" to avoid capture.[14] The RCC did so by providing safe haven for them in the Vatican while counterfeit birth certificates and passports were made by the Sicilian Mafia the Vatican had hired, and cash for their escape was given to the fugitives, including Klaus Barbie, the "Butcher of Lyons"; Adolf Eichmann, director of the genocide program; Eduard Roschmann, the "Butcher of Roga"; Dr. Joseph Mengele, the "angel of death" at Aushwitz;[15] and "Franz Stangl… the commandant of the notorious Sobibor and Treblinka death camps, where an estimated 1,000,000 to 1,250,000 Jews and Gypsies were gassed."[16] Nazi war criminals were sent by the Vatican mostly to South America, Australia, and the U.S., and many lived out long lives in freedom.

In the midst of WWII, on June 27, 1942, the Vatican Bank (VB) was created. With the addition of the VB, the Vatican gained an extremely secretive bank exempt from any investigations whatsoever. The VB "…is under the direct supervision of the pope. He is the one and only stockholder. He owns it; he controls it. Unlike any other financial institution, the Vatican Bank is audited by neither internal nor outside agencies."[17] In 1870 "Catholicism became the only religion in the world with the status of a country."[18] "The Pope was declared 'sacred and inviolable,' the equivalent of a monarch, but with divine right."[19] The pope is a man loved by a billion who is above the law, leads an untouchable country, and controls the most influential religion in the world with unlimited funds in a protected, secretive, international bank.

Annual financial reports now get published by the Vatican, but those reports have never contained anything having to do with the VB; the VB has always been completely separated from the rest of the RCC's financial disclosures. "Because of its clandestine workings, millions may be deposited into the Vatican Bank and disappeared into numbered Swiss bank accounts. This proved an ideal means of dealing with fraudulent securities, Mafia money, and Nazi gold."[20] The VB secreted away tons of Nazi gold, including gold that had been taken from the bodies of Nazi victims, forever remaining in irreproachable denial of its wrongdoings.[21] The VB destroyed all its banking records up to the year 2000,[22] records which it had never shared and hasn't to the present day.

Though the RCC found great fault with the Jewish religion, judging it for making money more important than following the directives of God, this is exactly what the RCC has done since its inception, most noticeably since its part in the development of WWII and its actions within the VB.

Mafia businessmen found the VB served them very well, since both operated in the dark, as criminal activities over time laundered and shifted billions of dollars through the VB that could never be traced. Paul Marcinkus, an American bishop from Chicago, rose to the position of president of the Vatican Bank in 1971 after years inside the Vatican. The VB president became the Mafia's inside man and was answerable to no one in the world other than the three consecutive popes he served under until 1990. It was directly through the VB that Marcinkus helped orchestrate the international, billion-dollar scandals of Michele Sindona[23], Roberto Calvi - both Mafia - and the Banco Ambrosiano affair[24], among others.

Because these scandals are all very well-documented in dozens of books, suffice it to say the Vatican's scandals involved a long list of the highest of crimes. In just one deal, the Vatican, through cardinal Tisserant, ordered $950 million in counterfeit American stock and bond certificates from the Mafia produced on American Mafia printing presses.[25] In other crimes, the VB was a major shareholder or stakeholder in dozens of banks and hundreds of shell companies throughout the western world that Marcinkus, Sindona, Calvi, and others had set up, many of them tax-exempt and phony "ecclesiastical corporations." Using the principles of fractional banking begun with deposits of counterfeit securities, billions of real-dollar loans were made to all these companies in an international racketeering Ponzi scheme. With the VB being immune to all external and internal investigations, billions of dollars in Mafia narcotics profits have also been laundered through the VB. The blatant murders of investigators and others when all these schemes finally began to fall apart, along with the complete noncooperation of the impervious Vatican, ended many of the investigations.[26,27] John Paul I threatened and began to make deep reforms within the VB and died under shadowy circumstances after only one month as pope in 1978.[28] Pope John Paul II then promoted Marcinkus to archbishop and "mayor of Vatican City… the third most powerful man in the Vatican, behind only the Pope and the cardinal secretary of state."[29]

The Godfather movies were even made by the Mafia – as were its profits - as Paramount Pictures was owned by Sindona's Immobilaire, a giant Italian company he had bought through Marcinkus and the Vatican.[30]

The RCC has operated in complete darkness for at least seventeen centuries, and it has only been three years since international banking reforms finally began to be forced upon the entirely clandestine Vatican Bank in response to the "war on terrorism." The RCC, the historic, centuries-old enemy of women, nature, and science, is just beginning to accept being accountable, like a snail tentatively raising one of its antennae in order to possibly move slowly forward, with the fear of exposure its greatest concern; that, and the resulting loss of revenue. Its banking reforms are supposedly going to be complete by 2018.[31]

All the illegal activities have been denied and covered up with as much religious and other immunity the Vatican can claim. The same has been done by the RCC regarding the child sex abuse scandals that have hounded the Church forever.

In modern times, the sex abuse scandals of the RCC only became public in 1984, but the first book written about this exact issue was in 1048, nine-hundred-sixty-eight years ago, titled *The Book of Gomorrah,* by St. Peter Damien[32], and "there is on record Church legislation going back to the fourth century concerning priests who sexually abuse people but especially children."[33] Like the other Church scandals, the modern sexual abuse scandal is well-documented. However, it is downplayed and referred to mostly as molestation or pedophilia instead of often being the actual rape of children in every orifice there is, often for years. The problem is, not much has ever been done about it in the RCC. The abuse is denied by the hierarchy and severely minimized, and guilty priests are merely sent to other parishes absent their criminal history, protected by the Church.[34] Such lawbreakers outside the Church would be thrown in prison or at the very least would be required to register as sex offenders for life. Rarely is a clergy member incarcerated. Although it's estimated only 20 percent of sexual-crime victims report the events, up to $4 billion has been spent by the RCC to settle claims just in the U.S. since 1984 and counting.

The former priest and author, Richard Sipe, reported that by 1985 the percentage of priests that were gay rose to about 40 percent, and that 6 percent of American priests "were either occasionally or regularly violating adolescents or children."[35] That adds up to several thousands of priests. In Ireland they say "the Church 'is the largest pedophile ring in the world…,' and Prime Minister Enda Kenny condemned 'the dysfunction, disconnection, elitism, and the narcissism that dominate the Vatican to this day.'"[36] It is said the final straw that caused Benedict XVI to be the first pope to resign in 600 years (in February, 2013) was the top secret, 300-page report given to him from three cardinals he had appointed to do the report, which "exposed in detail a 'gay network' of ranking clerics… [of] regular sex parties and that as a group, not only did they exert 'undue influence' in the Curia but that some of them were blackmailed by lay outsiders."[37]

Many say spirituality is a personal relationship with God, but religion is a governing hierarchy that claims to know what that personal relationship should be and dictates how everyone should live - a hierarchy that can't even live by its own rules it claims to have received directly from God. It's incredible that billions of people still choose to enable and "aid and abet" these institutions, supporting them to stay financially in the black – the black, as also in the shadows. Richard Sipe is quoted as saying, "The Catholic Church is basically a crooked institution. You know it. I know it. How is it that everybody else doesn't get it?"[38] The causes of the HOME are built upon ghastly illusions people still believe in, won't reevaluate or let go of, and even go into defensive rages over.

This is not about whether God exists or not; it's about what a criminal institution has done with its immense global influence. However, it's not all about the Catholic religion; it's just the biggest, richest, most independent and influential of them all. All the Abrahamic religions are the same: all teach to have as many children as possible, all teach that we're separate from and superior to nature, and all adamantly profess to be the moral authorities of the world.

That said, the RCC has had direct causation in the three most malevolent holocausts in history: nine million people burned at the stake during the Inquisition, tens of millions of Native Americans savagely murdered on two continents during the Age of Discovery after the pope gave them to Spain and Portugal, and the millions of "undesirables" horrifically exterminated during WWII. Along with its long history of Mafia involvement, its profuse cases of sexual crimes perpetrated upon minors and more, the Vatican still claims to be *the* moral authority of the world.

A gigantic Catholic following goes along with it, denying it all by looking the other way, thinking it's just made up by "enemies of the Church" and can't be real, or they find the Church and its culture somehow gives them comfort according to their normalcy bias, so they continue to support the Church with untold millions in "Peter's Pence" donations. These are not dissimilar to donations invented by the Church in the sixth century when they started the selling of indulgences, "whereby the faithful paid for a piece of paper that promised God would forgo any earthly punishment for the buyer's sins. ...The Vatican set prices according to the severity of the sin."[39] Later, indulgences helped pay for the Crusades. Still later, indulgences were applied whereby "[a]ny Catholic could pay so that souls trapped in Purgatory could get a fast track to Heaven. The assurance that money alone could [do this]... was [so] powerful... that many families sent their life savings to Rome. So much money flooded to [Pope Sextus IV] that he was able to build the Sistine Chapel."[40] By the early 1500s, Pope Leo took the sale of indulgences one step further by making payments "available for sins not yet committed. So much cash flooded in that he could build St. Peter's cathedral."[41] "Indulgences" would be called fraud or extortion anywhere else. The Church had peoples' confidence and used it to its advantage, the very definition of con artistry.

The point is, combined with the RCC's historic anti-birth control policy, its criminal activities, and the dominant ideology it uses to keep the world in the dark, the new Holocaust we're all facing is the human-caused ongoing mass extinction of species. All biota is suffering an ongoing anthropogenic Holocaust instigated in very large part to Abrahamic (and other) religions and the fantasy-bonded people who unquestioningly support such very destructive ideology and behavior with blind devotion.

BIRDS

F OSSILS SHOW THAT BIRDS developed with feathers at least 50 million years before the end of the 165-million-year Age of Dinosaurs and shared many evolutionary traits with dinosaurs. Both nested in similar fashion and had specific types of lungs, hearts, and skeletons in common and more.

About 2,000 birds went extinct with the colonization of humans in the Pacific[1], so we know there were previously around 11,000 bird species in the world. Of the approximate 9,000 bird species left today, according to the International Council for Bird Preservation (ICBP), almost one out of four birds in the world are currently endangered or threatened with extinction. The still-critically endangered California condor was recently just barely saved from extinction; over one hundred years ago it had a range along the Pacific Coast of North America from British Columbia to Baja, Mexico.

The threats to the survival of songbirds (as with all others) are many: deforestation, transformation of grasslands to farming, drainage of wetlands, the pet trade, introduced pests, invasive bird species, chemical pollution in the food chain, and climate change.[2] Grassland species are experiencing more intense population drops than any other bird group in the U.S.[3] Most exasperating is all countries around the Mediterranean Sea are devastating their gorgeous songbird populations with sport hunting and human consumption (even though most of those birds consist of only two bites of food), bringing disaster to these beautifully singing birds that migrate from Europe, Asia, and Africa to reproduce. Now they don't even have a safe place to land anywhere in the Mediterranean.[4,5]

Hummingbirds (which are unable to walk or hop) are gravely

endangered throughout Central America with extensive habitat converted to growing coffee and marijuana.[6] One critically endangered hummingbird, the Honduran emerald, is confined to an area just 4.5 miles square, its former habitat destroyed by raising cattle, pineapple, and rice.

"Hundreds of thousands of wild birds are illegally smuggled every year... approximately 250,000 [are] illegally shipped into the United States [alone]."[7] The illegal capture and trade of wild, endangered bird species is drastically threatening their survival. "Parrots are the most threatened group of birds in the world. Every year hundreds of thousands of parrots are caught for the caged bird trade, the majority illegally, and many die before being put up for sale. Combined with rapid deforestation, this unsustainable pressure has put more than one-third of all parrot species at risk. ...Nineteen species have died out in historical times, while a few dozen more are on the verge of extinction."[8]

FARMING

THE VIRGIN, UNBROKEN PRAIRIES of America were once teeming with species equal to that of the African savannahs. Millions of acres of thick interconnected root systems crossed the continent, thriving in the richest soil on Earth. In some places, the topsoil was one hundred feet deep, developed over hundreds of thousands of years at an average rate of one inch per 500 years. Upon this grew a dense array of 150 wildflowers and forbs (non-woody plants other than grasses) that displayed unimaginable colors along with another 150 species of grasses, some of them nine feet tall such as big bluestem.[1] Feasting upon this paradise of food across North America were vast herds of wild ungulates. After the wooly mammoths, saber-tooth tigers, camels, and the first horses, tens of millions of antelope, pronghorn sheep, elk, caribou, bison, and many others roamed the continent along with their many feline, ursine, and canine predators. There were billions of prairie dogs, millions of beaver, turkeys, foxes, grouse, and owls, to name a scant few.

The unravelling of the American savannah began in earnest by the early 1830s with the arrival of the European settlers, their religious ideology, and their polished steel plow. "The plow cut through tough rootlets and plant spurs with a sound like fusillades of tiny pistol fire, all amplified by the tempered steel moldboard in a steady ringing hum that might last fourteen hours a day. In the wake of the plow was the confusion of the prairie's dispossessed...."[2]

The continent was shredded apart for decades, crisscrossed in lines and squares. With the former network of roots gone, the soil dried out and billions of tons of soil were washed away or blown out in giant dust storms. Contour plowing wasn't even thought of until after the

Dust Bowl of the 1930s and only begrudgingly applied. The richness of the soil allowed for these losses, and America established itself as "the breadbasket of the world" even to this day. A wealth of nature that preceded the onslaught of agriculture was annihilated, along with untold centinelan species, "buried forever by corn and commerce,"[3] though humans have benefitted tremendously. However, even now "[s]ome 25 billion tons of topsoil are... being lost each year with untold consequences to the food supply...."[4]

Nitrogen, abundant in the atmosphere, is a vital nutrient to plant growth, but plants can essentially only get it from the soil. By 1913, a man named Fritz Haber invented a process to extract nitrogen in great quantities from the air and apply it to the ground as fertilizer; it revolutionized farming and increased crop yields tremendously, especially in the U.S. Applied in the millions of tons, "[i]ntensive agriculture currently uses more nitrogenous fertilizers than the total amount of nitrogen naturally fixed by all of the earth's ecosystems."[5]

Beginning in the 1940s, while it exported crops to many countries, America initiated its "green revolution" in an effort to prevent probable starvation in the world, sharing technologies with countries around the world to improve their farming. Though originating in the Yaqui River valley of northwestern Mexico, by the 1960s the green revolution took hold globally. The Green Revolution allowed the world's population to double between the 1960s and 1990s, creating a vast amount of poor people around the world, as they were driven from their farmlands by mechanization.[6] "[I]t was possibly 'the worst thing that has ever happened to the planet.'"[7] Meanwhile, by the 1990s, the Yaqui River valley dried up from farming abuse and was largely abandoned. In near proximity, the same is true for the lower 300 miles of the Rio Grande, now dominated by a very tough weed species – tamarisk, also known as salt cedar - which can withstand fire, drought, searing heat, and floods.[8]

America has surpassed all other countries in farming for a century. For example, with ever-increasing farming technology, the U.S. averages seven tons of corn per hectare, compared to a worldwide average of four tons, and has gotten the yield up to twenty-one tons in Iowa.[9] In the early 1980s, the U.S. stepped up its agricultural economic efficiency under the Reagan administration by adopting new farming policies to support planting crops from fencerow to fencerow. Since then, for as far as the eye can see are fields of monoculture crops, many now genetically modified, with scarcely a weed in sight, further decreasing wildlife drastically.

It requires 130 gallons of water to grow a pound of wheat, sixty-five for a pound of potatoes, 265 for a glass of milk, 530 for a pork chop, 3,000 for the feed to create a quarter pound of hamburger, thirty-seven for a cup of coffee, and three gallons of water to grow a teaspoon of sugar for that coffee.[10] When finished agricultural products are exported (or imported), it is known as the virtual water trade, which refers to the unseen amount of water attached to growing those products. The U.S. and Canada are the world's largest exporters of virtual water. Mexico, the Middle East, North Africa, Japan, and much of Europe are net importers. About 40 percent of all human water consumption is rooted in the worldwide trade of agricultural products.[11]

About 40 percent of Earth's land surface is devoted to farming. All land that can be converted to farmland has already taken place unless more rainforests are destroyed. Additionally, erosion destroys up to 20 million acres of cropland per year, and a net 5 percent of land devoted to raising grains has lost its production since the 1980s. Just four crops – rice, wheat, potatoes, and corn – make up over 50 percent of farmed produce, and only sixteen more crops provide a total of 85 percent of food. "[O]nly fifteen mammal and bird species supply 90 percent of global domestic livestock production"[12], seventy percent of available freshwater goes to agriculture, salinization wipes out millions of productive acres per year, desertification claims about 15 million additional acres annually, about one-third of crop yields are reduced by pests and disease, global aquifers are being used up much faster than they can be replenished, and over 90 percent of global crop species have vanished in the past one hundred years. Of all global food production, about 75 percent comes from only a dozen crops, 85 percent from twenty, and 95 percent from thirty, though humans historically relied on consuming up to 5,000 plant species. Since 1900, more than 6,000 varieties of apples (86 percent) have become extinct in the U.S., 91 percent of corn varieties have disappeared, as well as "95 percent of cabbages, 94 percent of peas, and... 81 percent of tomatoes."[13] By 2006, the global area taken by GMO crops outnumbered organic crops by 400 percent[14]; this does not include conventionally grown crops. Farms are having their water rights legislated away to shift more water use to cities, as the global urban population reached 51 percent for the first time in history in 2013. Moreover, 65 percent of bird species are threatened with extinction by agriculture.[15]

Farming does not operate in a free market; instead, farming is heavily subsidized, given tax breaks, and the market is further manipulated with tariffs. When water use is not subsidized, for example, more appropriate use flows to crops with higher value that use less water, and conservation is encouraged. Rather than broadly spraying in the middle of the day, which maximizes evaporation, systems would be used that bring water directly to the roots, and more irrigation might be done at night. However, water and its subsidies are legislated, and legislation is a very slow process, generally far too traditional and resistant to make any necessary, forward-thinking changes any time soon; legislation is dominated by crisis management rather than fundamental, innovative problem solving. Meanwhile, much of the world is drying up and mimicking the desertification of the Rio Grande.

Furthermore, vast amounts of fossil fuel are used for industrial farming and its global supply chain: "[n]atural gas in the fertilizers, petroleum-based pesticides, fuel for all the tractors, more fuel to transport the food to processors, fuel to process the raw crops into food additives, then to manufacture them into products, and then to transport the products across the country or world… [and shipped] thirteen hundred miles on average [to the consumer]…."[16] We're using oil and the "principle" of aquifers to temporarily boost farming production all for the DIC's demand for economic growth. On top of that, it only works because all these costs are transferred onto taxpayers through subsidies that prop up agribusiness, along with taxes that fund the military to secure fossil fuel. Our system socializes costs that allow MNEs to make private, multibillion-dollar profits, and then the MNEs dictate the control of our food supply and our foreign policy of never-ending wars to secure fossil fuels.

All the while, the DC never gives a thought about how it's causing the mass extinction of species.

NATURAL SERVICES

H UMANS ARE CURRENTLY TAMPERING with the natural services of the world. The unrecognized value of annual services freely provided by nature surpass the yearly $30 trillion global GDP of humans. These services include the pollination of crops, soil fertility, medicine, recycling of wastes, purification and storage of fresh water, food, building materials, photosynthesis, carbon absorption, watering of crops, cooling and heating of air, creation of topsoil, energy (solar, electromagnetic, wind, tidal, gravity, fossil fuels, geothermal, and others), circulation of oceanic nutrients and temperatures, protection from ultraviolet light, minerals, and more. As we alter and erode all of these systems, we endanger all of life as we know it, and we are doing so at an increasingly rapid pace. For instance, without insects, terrestrial ecosystems would fail; without bees, the best one-third of our food would be gone. Along with that, we've caused the loss of 20 percent of both Earth's topsoil and agricultural lands and about 90 percent of commercial marine fisheries[1], not to mention forests, in the past sixty years.

Terrestrial ecosystems generate "about 132 billion tons of new organic matter each year" globally.[2] This does not include the organic mass of similar production in the ocean or of all animal growth. Humans now consume or destroy a majority of these organic resources, having an enormous impact upon ecosystems and other species. Humans are already trying to compensate for some of these freely-given services with artificial means, such as pollinating crops by hand, as they do in parts of China, or building desalination refineries, or fertilizing crops with manmade chemicals, or creating energy at nuclear fission sites.

The altering of natural services is the result of seeing nature as a mechanical thing full of commodities; however, the biggest error in the DC's mental illness is in not seeing or valuing nature for its priceless amenities. "Psychologists have in fact discovered that just a view of natural environments... leads to a decline in moods of fear and anger, and it generates an overall feeling of tranquility."[3] Humans have inherently evolved as a direct result of nature, which has had 3.4 billion years of genetic trial-and-errors, culminating in the successful manifestations of natural selection. As we destroy the life of our planet, we destroy ourselves, our feelings of well-being, and our peace of mind, spirit, and soul; we increase our fear, unease, anger, and illness on all levels of our being. To experience a healthy Earth maintains our sanity: to be in the quiet of a sacred, ancient forest; to have a vista of untouched land for as far as the eye can see; to at least psychically bond with gorgeous, innocent animals in the wild; to breathe pure, fresh air and drink free, pure, untreated water; to see thousands of migrating birds in the sky and millions of other beautiful spectacles.... All this gives us a sense of sanity that modern humans no longer have a point of reference for in their psyches. Instead, if they're not too distracted by their consumerism and technological gadgets, most at least unconsciously feel our Earth being continually degraded, imbalanced to the tipping point of utter devastation or complete artificiality at best. Most are unable and unwilling to see through their normalcy biases and denials that our world is swiftly beginning to sink into the black hole of our human-caused mass extinction of species with our disruption of the Earth's natural services.

POLLINATORS

IN ORDER TO FLOURISH on land, DNA had to learn how to make use of terrestrial Earth's abundant fertility after photosynthetic eukaryotes were lapped upon the shore for eons with the ebb and flow of tides and fluctuating wetlands. Fungi helped plants take hold on land, evolving into completely new forms of life. Spores were the first method of reproduction for the two symbiotic kingdoms and included mushrooms, mosses, ferns, and lycopods, which evolved into the first trees. Lycopods were so prolific that they now comprise the majority of the world's coal deposits; eventually their abundance emitted so many flammable spores into the air that the forests ignited in self-destruction about 280 million BP.[1] Lycopods were replaced by gymnosperm (exposed seed) forests, such as redwoods, which emitted pollen spread by the wind - pollinated the same as grasses, including wheat, rice, and corn today. Through these steps, it took 1.17 GY (from 420 million BP to 140 million BP) for the Kingdom of Plantae to develop into the complexity of angiosperms (flowering plants). Flowering plants have ovaries (hidden seeds); the pollen has to travel up pistils to fertilize the seeds, creating embryos. Until angiosperms, there was no need for pollinators.

In the age of dinosaurs, arthropods first developed into terrestrial insects, then birds evolved, then angiosperms, all within about 50 million years of each other. Insects have especially had a symbiotic relationship with plants since their inception; it's likely they were instrumental in plants evolving into angiosperms. With the extension of filaments (pistils and stamens) to produce and receive pollen, and the pollinators to spread the pollen, angiosperms became more specialized than reliance on the wind, and life went on to spread into more diverse, particular nooks and crannies of habitat. Bees, in turn, developed a flabellum, a tongue with a fringed, spoon-like flap at the tip that allows it to rapidly lap up hidden nectar within each flower.[2] Angiosperms and pollinators have had spectacularly successful symbiotic relationships ever since: eighty percent of terrestrial plant species are flowering plants.

The first pollinators of early angiosperms were "beetles, thrips... moths, and flies."[3] The estimated number of pollinator species is now over 130,000. They include about 25,000 types of bees that pollinate 73 percent of plants along with flies (19 percent), bats (6.5 percent), wasps (5 percent), beetles (5 percent), birds (4 percent), and butterflies and moths, (4 percent).[4] Other species that contribute to this overlapping service are opossums, monkeys, lemurs, mosquitoes, geckos, some marsupials, "tree squirrels, bush rats, galagos, tree shrews, raccoons, kinkajous, olingos, and long-tailed weasels... [and] bird species in at least 18 families have been confirmed as effective pollinators of plants."[5] Hundreds of these pollinating birds go by names such as yellow-footed honeyglides, four-colored flowerpeckers, mahogany gliders, and turquoise-throated pufflegs. Nearly *two-thirds* (about 85,000 species) of floral pollinators now include threatened species.[6]

On both sides of the equator, bees greatly increase in numbers and species outside the tropical zone; they are most prolific in the deserts and savannas. Pollination of the tropics is left to the many other pollinator species. Once outside the tropics, pollinators decrease in numbers through the temperate zones and on towards the Poles. For instance, in 1996 Mexico had 471 known species of butterflies with forty-six of them endemic, the U.S. had about one-third less with half as many endemic, and Canada had about two-thirds less with only two endemic. This pattern is the same throughout the world in both hemispheres and parallels that of bee species, as "all the... 160 genera of bees known from the North American continent can be found in Mexico."[7]

All pollinators are not equal, especially when it comes to food crops. This is especially pertinent since 80 percent of global plant-food species rely on pollination. Pollination also provides food for our domesticated ungulates and clothing for us; it is a natural service with a value of over $16 billion a year just in the U.S. Some pollinators, such as honeybees, are generalists; most others are specialists that only have relationships with a certain narrow range of angiosperms that offer specific aromas, colors, forms, and more. A decrease of pollinators leads to a reduction and even extinction of plants. Flowering plants do everything they possibly can to attract pollinators, indicating that pollinators are in high demand even with healthy populations.

More than any other species, honeybees are used worldwide for the pollination of our food crops. Honeybees are an introduced, domesticated species. Of course there are some that have gone feral, and Africanized honeybees are a wild, invasive species; both types of honeybees are weed species and disturbance-loving species that follow in the wake of human destruction of the natural environment. Chosen for its docility and easy handling, the prolific *Apis mellifera* honeybee is inferior to but displaces many native pollinators. It is incapable of buzz pollination, which many food crops need; for example, the buzz-pollinating bumblebee actually vibrates a flower with a certain pitch that's necessary to set the pollen for tomatoes, wildflowers, and certain fruits. Many biologists refer to honeybees as "lilliputian livestock – fuzzy herbivores with wings – that are just as capable of taming a landscape as any cattle, sheep, or goat infestation. Their 'grazing' on pollen and nectar simply goes unnoticed."[8] "Moreover, they pack their pollen wet with nectar and saliva – thus rendering it unviable, or at the very least not so likely to be rubbed off and donated to a waiting stigma. ... this makes honeybees 'ugly' pollinators...."[9] This being said, we need honeybees for the world we've created because of their traits as a weed species, yet they have been in steady, drastic decline over the years, starting in the 1970s. But now, it has reached a frightening condition referred to as Colony Collapse Disorder (CCD), which is global.

Annual declines in the honeybee population are everywhere and have reached 70 percent in many areas on six of our Earth's continents. There are many causes for this: habitat destruction and fragmentation; heavy use, ignorant overuse, and flagrant misuse of agricultural chemicals; fence to fence crop planting, leaving no room for hedgerows; microwave radiation from cell phone towers and smart meters; high-tech, laser-leveled fields that become too sterilized; invasion of Africanized honeybees; two fatal mite species; and certain fungi, bacteria, and viruses bees formerly had more immunity to.

The courageous and brilliant Rachael Carson proved correct over and over in a sea of misogyny when she said things such as, "A bee may carry poisonous nectar back to its hive and presently produce poisonous honey." (This principle may also apply to why 95 to 100 percent of the populations of free-tailed bats have died from half their caves in Mexico.[10] Bats are also dying by the millions across the U.S. Perhaps it is their dealing with all the poisons in the environment from the insects they eat that have weakened their immune systems against the white-nose fungal disease. Ten bat species are already extinct.) Rachael also said, "Now clean cultivation and the chemical destruction of hedgerows and weeds are eliminating the last sanctuaries of these pollinating insects and breaking the threads that bind life to life." Still, such commonsense, scientific insight only temporarily stalled the DC's juggernaut, the blind devotion to the steamrolling ways of the cultural status quo.

Mexico's use of pesticides went up over 300 percent in the eleven years before NAFTA and increased even more from there; plus, like many other countries, they still use chemicals that are known to kill bees and are banned or controlled in the U.S., chemicals such as DDT, heptachlor, malathion, aldrin, and chlordane, to name a few. Similarly, "Brazil is both the major user and the major producer of pesticides in Latin America."[11]

The steadily increasing occurrence in the loss of bees since the 1970s includes the associated waning of plant production. This "clearly applies to rare plants that cannot sustain pollinators or seed dispersers necessary for the regeneration of their populations."[12] But the DC has the answer of course – they have developed artificial bees, "robobees," drones shaped and sized like bees without stingers. Rather than rectify any problems at their fundamental level, a new industry will once again save the day while also boosting and not disrupting the economy. Once again, humans celebrate their mythic illusion of being better and bigger than nature. That way we can even increase our use of chemicals and destroy more habitat to feed more people. Why change when we can self-pleasure ourselves with technology into sustained delusion? Every problem has a glorious opportunity, after all. Who needs bees? Who needs nature? Think of the money to be made manufacturing and selling artificial bees! All it shows is how far the DC has removed itself from reality. The DC has actually made several versions of these artificial, invasive species, so don't worry - *they're only replacing natural pollinators!*

DEBT

HUMANS HAVE CREATED AN environmental debt, which includes the carbon debt, the radioactivity debt, the chemical debt and others, all culminating in the extinction debt. This has happened as a result of what we have done to the environment, which will take decades to show its full effect. Atmospheric carbon dioxide is a good example; scientists figure it takes forty years for the effect of its presence to fully manifest. Like that coal-fired freight train, the impacts to the Earth and its species we have set in motion do not have brakes. How do we put brakes on carbon already in the air or from Polar ice melting? And we can only put endangered species in zoos or store their DNA in frozen vials in desperate attempts to delay the extinction debt.

When the wolves are gone, which allows the elk to take over and eat all the tender, young aspen shoots, the aspen forest enters into an environmental debt; the aspens in those areas become a ghost species, ready to die out when those still standing finally pass and have no younger generations to replace them.

Such lapses of time give plenty of fuel to the deniers who say there is no problem. The debt that has yet to catch up also gives plenty of time for the humanists to think they can come up with the solutions to any crisis. The debt takes long enough that new generations of people are born with a new normalcy bias. The new Ark - Neo's Ark - sinks in denial until it's too late to save it because no one wanted to incur the unrecoverable expense of repair and restoration and lose their business in the competitive economic market if they were to do so. Businesses that are in power will ride their wave as far as it takes them through the debt because they will not voluntarily lose profits when the debt can be made to seem vague and unprovable, when the time lapse it takes can't be specifically pinpointed to any particular company, or the companies have enough power to be able to fend off any blame. The debt takes long enough to manifest that those warning the world can be ridiculed and shown to be alarmists with faulty predictions, strengthening the denial. People hear and believe what they want to believe because they don't want to have to change, and they love to have their fears abated by their chosen "experts."

How much is in the Earth's savings account of natural resources? At this point, we are living off all the interest – all that the Earth can produce in a year – plus up to another 50 percent and rising. Every year we are drawing off the Earth's principle by using up to 50 percent more than our Earth can generate. The freight train is yet still gaining speed and consuming much more than is sustainable in the long run and is exponentially spewing out more toxins. When will it be too late to stop the train no matter what the humanists come up with – when all the tigers, rhinos, tropical forests, orangutans, whales, wolves, polar bears, tuna, gorillas, gibbons, chimpanzees, salmon, halibut, bats, a plethora of plants, pollinators, and a million other species are all gone? It's getting very late, even if we change *now*, but especially if we do not. The environmental and extinction debts are not meaningless concepts fabricated with electronic digits or paper, the intangible notions of humans only cognizant of their own fantasy worlds.

Even if we completely stopped everything we're doing wrong to the habitats of Earth and ultimately to its species, the ozone layer will take a century to heal; the CO_2 we've added to the atmosphere will remain for more than a century and continue to warm the climate; the pollution of the oceans and closed seas will infiltrate into the food chain, especially of radiation and miniscule, decomposed plastic that will last for millennia; eroded soils will take several centuries to be restored; exploited populations of fish will require several decades to regenerate, not to mention the coral; and areas we've turned into deserts will last for innumerable decades or more.[1] The Earth is being steadily and increasingly depleted, exponentially depleted, even while the environmental debt is accumulating.

NATIVE AMERICANS

THE NATIVE AMERICAN CULTURE, in broad terms, has virtually been ignored, and this ignorance has had vast consequences upon our world. The world has been under the dysfunctional ideology of the DC for thousands of years, yet for over 500 years there has been an entirely different mindset to learn from. The self-righteous DIC, however, has been incapable and averse to doing this now-crucial act of humility.

We know the DC mythology starts with humans having been created separate from all other life by a male God, that humans got kicked out of an imagined paradise because of Eve, and that humans have been punished ever since, left to live in shame, guilt, blame (of women), and "sweat of the brow." We may not all believe it personally, but it is the cultural story that underlies why things are the way they are. The story tells us that it was ok to take two continents of Native Americans' land, which *was* a paradise, because the DCI judged they were not "putting the land to good use." The DIC then put paradise to "good use" by destroying it to increase its population and gain cultural, religious, and personal power and profit, allowable because they said God gave humans – the DC in particular - the exclusive right to take anything they wanted under the semblance of dominion. To the DI, nature has been an enemy to subdue and conquer, full of problems which must be eradicated, a place where evil lived in the form of terrifying animals and superstitious misfortunes. The DC has never wanted to live with nature but has rather wanted to overcome it. However, nature was also full of riches, and to the DIC wealth has always been recognized by how much one could gain, take, steal, and keep; greed is accepted as something good, something to strive for, or at least considered normal as a way to "get ahead" in an individualistic (fractured) culture. The DI chooses to believe that the Earth was made for man because its members are simply too greedy, self-important, and disassociated from nature (by preference) to believe otherwise. As an individualistic culture, throughout history the DIC has used the guise of honor as a trick to do dishonorable things, where any means justified the ends, where the honorable were often the naïve, trusting losers. This is not only ancient history: rapacious MNEs are the epitome and current manifestation of the DC's dishonorable tactics, including the mindless apathy and self-absorption that's sending millions of species into extinction. The DCI is like the bulldozer it invented and has used it by whatever means necessary to perpetually get its way because, through such power, it got to say it was right and do whatever it wanted, including saying that its actions have been ordained by (its) God.

It didn't have to be this way, and it still doesn't, but a radical change needs to take place within the realm of humans. Traditional Native American ideology provides a different model to live by, with precepts that could be applied to our modern world. The world cannot and never does go backwards; adopting parts of Native American ideology isn't about going back to teepees and buffalo hunting as those of cataleptic mind immediately imagine. It's about injecting sanity into the DIC paradigm. Finally embracing and integrating Native ideology is vital in order to build a functional new paradigm.

If we turned the DC's ideology around to the exact opposite, we would have Native American ideology. Traditional Native ideology begins with creation stories of most tribes emerging directly from the Earth, the Mother, even from certain physical openings. This metaphorical birth from the Earth Mother not only makes sense but has also been scientifically proven through genome mapping; the exact same basic DNA sequences exist within all lifeforms on Earth, proving we are all related and developed from the Earth. Natives were living in an actual paradise, never got kicked out to live in shame or sin forever, and paradise didn't need changing or improvement. Native Americans lived at one with the wilderness and didn't even know it as such. Without a directive to "be fruitful and multiply," they knew the carrying capacity of their environment and consciously limited their populations in order to maintain the balance of life their own lives depended upon. Harmony with the environment was their ultimate goal, not increasing their population no matter what; they were not trying to dogmatically overtake others by sheer numbers. By preserving the wilderness intact, Natives were maintaining the best use of the land. Native American ideology sees we're all part of the interconnected nature of the Earth, our sustainer. All plants and animals are part of us, which is true, and, as all one interrelated family, Natives knew they must honor, respect, and take care of all their relations, to live symbiotically in mutual agreement. They never saw nature as an enemy, and it was not full of Satan manifesting as evil beasts. Natives have always known humans belong to the Earth and have the responsibility of taking care of Her, leaving as little impact as possible, knowing that what we do to the Earth we do to ourselves because we *are* Her. Natural law ruled, and Natives didn't try to "conquer" it, for they knew they were part of it, not separate and above it with some false understanding of dominion given to them from "on high" to do with as they wished. Native American culture historically was the most generous on Earth and demonstrated an absence of greed; in fact, wealth was recognized by how much one could give. As a tightly collectivist culture, Natives lived lives of honor because they had to apply what worked in the real world; for self-preservation of the tribe, the dishonorable were dealt with so harshly that such occurrences rarely happened.

The DCI came to the Americas and judged the Natives as uncivilized, yet what has civilization been but a system founded upon domination, upon conquering, which is fundamentally uncooperative. In particular, what was the conquering of land but a dysfunctional concept from the start? Only the DC would think of land as a foe. It's like the armies of the insane Roman Emperor, Caligula, being ordered to wade into the British surf to slash and stab the ocean waves to conquer King Neptune, returning to Rome with chests of seashells as booty to prove they won. Nature does not fight back and is not at war with us. Land doesn't fight back, so it was never conquered but only by a misguided concept. It was the Treasury of Life that was raided and stolen. Native Americans were the guardians of the treasury; the guardians were overwhelmed - conquered through disease, guile, superior weaponry, and an endless supply of newcomers - and the treasury stolen. The Indians have always said the DC is destroying its own home, and even now that all the frontiers are gone and all the land is "conquered," the DI mindset remains the same. Native Americans never even thought of the land as something to conquer; the land was something to love.

"Any person having witnessed and tasted the life of the savage would not dream of returning to white society." - Benjamin Franklin.[1] The uncooperative DC civilization has been full of endless wars; constant, severe butchering of other humans; conscious treachery and dishonor; murder of family members within aristocracies to manipulate, usurp, or control greedy agendas of power; burnings at the stake and gas chambers that killed millions and nuclear holocaust which threatens more; and, ultimately, of mindless extinction of species and ruination of entire ecosystems.

Unscrupulous people of the DCI have been the "civilized conquerors" - an oxymoron - and the conquerors got to crown themselves as civilized, thereby determining their needs as primary over the conquered. With a bias that the Indians were an undeserving and inferior race, the DIC saw the Americas as an untouched, unused paradise – a bank account of natural resources (power) - and the continents were looked upon with avarice. From popes to presidents to settlers and now cultures around the world, the DI felt it had the right and the duty to steal Indian land because they didn't see the Indians using the land according to DC standards, whereas the Indians were shocked and horrified by how grievously the DC abused the land - which it continues to do.

"When Indians alone cared for the American earth, this continent was clothed in a green robe of forests, unbroken grasslands, and useful desert... filled with an abundance of wildlife. ...For all American Indians, ecology was not a separate subject or something to be concerned about only part of the time; it was involved in an entire way of life."[2] "Native Americans viewed themselves in relationship to the earth. They saw themselves so much belonging to the earth that... the earth was their mother. She sustained them and they adored her."[3]

"To the pioneers, Indians didn't count. By the criteria of the pioneers, the forest [or any other desired ecosystem] was unclaimed wilderness because it had not been cleared,"[4] so they also cleared the Natives out.

The DI civilization has been and is currently doing the same thing, most notably in the Amazon. "In 1964, there were about 250 indigenous tribes thought to be living beyond contact with modern society,"[5] but "[i]n... the twentieth century, more than 90 tribes have become 'extinct' in Brazil alone."[6] For instance, "[g]unmen [in 1963] allegedly hired by rubber plantation owners machine-gunned a Cinta Larga village located on the Aripuana River, killing thirty-seven hundred of the five thousand Indians," known as the "Parallel II Massacre."[7] The world hates all that happened in the 1800s to tribes in North America and thinks the murdering of Natives all ended then and there, but it's now even far worse for the world to allow such crimes to exist in the Amazon to this day. These days, tribal members are still murdered with impunity by members of the DC sometimes tying them to trees and killing them with chainsaws.

Another example of "civilization" was the ploy to steal Indian land, even after the reservation system was established, by dividing reservations into farming plots, and millions of acres of arable land were then given to non-Indian settlers. It was the various Christian churches that were the chief advocates of this policy, and they were the ones who rushed in to grab the choice allotments before the settlers[8], claiming free land under the Homestead Act and tax free forever under church privileges.

Yet another act of "civilized people" was to introduce epidemic diseases "deliberately, as when blankets that had been used by smallpox patients were given by soldiers to the Pawnees...."[9] European diseases alone killed up to 85 percent of all Natives of the Americas, impoverishing Native civilizations immensely.

Native American civilization was perceived as uncivilized in spite of highly advanced Incan, Mayan, Pueblo, and Aztec civilizations or empires with the largest pyramids on Earth (Teotihuacan) and bigger, more sophisticated cities than the Old World ever knew. They also had more exact calendars and greater understanding of astronomy than the Europeans at the time of the Americas' discovery and had more advanced artwork than the ancient Egyptian civilization, including the works of their goldsmiths. The Spanish, however, melted and stole all the gold and burned the libraries of the New World (as happened to the great libraries of Alexandria) to enrich themselves and destroy any evidence that could reveal that the Catholic Church's domination was solely based upon erroneous superstition. Anything in the way of gaining access to the new natural resources that could greatly enrich the avaricious DIC was eradicated.

Examples of the DI mindset towards Native Americans, from the most learned and highly esteemed leaders of America are quoted as saying, "Established in the midst of another and a superior race, and without appreciating the causes of their inferiority... they must necessarily yield to the force of circumstances and ere long disappear." - President Andrew Jackson. "This unfortunate race... have by their... ferocious barbarities [in defending the Treasury of Life] justified [their own] extermination...." - President Thomas Jefferson. "Discovery [of America by Europeans] gave an exclusive right to extinguish the Indian title of occupancy, either by purchase or by conquest." - Chief Justice John Marshall. "The settler and pioneer have at bottom had justice on their side; this great continent could not have been kept as nothing but a game preserve for squalid savages." - President Theodore Roosevelt.[10] These show the typical either/or, judgmental mentality of the chauvinist mind. These leaders were incapable of even thinking of the possibility of understanding, learning from, and cooperating with Native culture.

The DC has always believed it was so absolutely right about all things that it wasn't until *four and a half centuries* after DC discovery of the Americas that Aldo Leopold realized, in 1949, some of what Natives have known forever and at first tried to tell individuals within the DCI, but sixty-seven years later, the DI still hasn't changed: Aldo's ideas were just an anomaly, a novelty, meaningful only to an insignificant minority of ecologists. Mr. Leopold wrote: "We abuse land because we regard it as a commodity belonging to us. When we begin to see land as a community to which we belong, we may begin to use it with love and respect."[11] He also wrote a sub-chapter titled "Thinking Like a Mountain," alluding to how the DIC has failed to see the importance of working with the functions of nature for the long-term betterment of the environment and, hence, ourselves.[12] Native Americans have been saying this forever about keeping the Earth healthy for the Seventh Generation to come, that the unborn have a right to an Earth that we leave in best condition for all time. Leopold's sub-chapter could have been titled, "Thinking Like a Native American," but perhaps even he couldn't admit or conceive of the idea that Native Americans knew more than the DCI.

Jane Goodall, with Marc Beckoff, wrote, *in 2002*: "We are beginning to learn that each animal has a life and a place and role in this world."[13] How disheartening that only fourteen years ago, a researcher of the DC held in such high regard states that they are *just beginning* to understand such a fundamental concept that Native Americans have known and lived by forever. This indicates how much the DIC has needed to drop its arrogance and learn from Native American culture and how incredibly unconscious the DI mindset continues to be. And for all the extraordinary work Jane Goodall has done, she was ridiculed by the patriarchy for most of her career, even for such basic principles that animals have feelings and have important roles in the world. And yet, has the DC even listened to the likes of her? Has it even slowed down the pace of chimpanzees and other primates heading ever closer to extinction?

Another example: "We now know [1993] that the current industrial forest's... practices are wrong, and even some mainstream forestry organizations are seeing this truth."[14] It only took the DC 501 years to figure that out but only after they took everything they could with the loss of all the Ancient Forests! And, even knowing, they still haven't changed their forestry "practices."

The point is, the DIC has at best only begun to see the errors of their ways, has not changed, and may not, even when the ongoing mass extinction they've caused is in full bloom. Native Americans are not in the least impressed by such "groundbreaking realizations" of even the most outstanding voices of DI naturalists; even after twenty-one generations (524 years), Natives are still dismayed that the DCI continues to be so completely and absolutely ignorant and uncaring about how to live in harmony with our environment. Traditional Native Americans would never have built dams that killed precious salmon runs and never would've plundered entire forests, for example, but the DC came to force the land into supporting its way of life whether the land was matched for it or not. Dust Bowls, desertification, loss of aquifers and riparian ecosystems and more are the result.

The DC is always trying to vindicate its deplorable environmental conduct by citing alleged misdeeds Natives might have done up to ten thousand years ago in the vestiges of the Stone Age similar to running herds of bison over cliffs. (One such site exists near Casper, Wyoming from 10,000 BP). Nonetheless, over 60 million bison existed when the DIC came to the Americas, but the DC wiped them out to near extinction in just a few years. Any comparisons that the Natives were no better than the DCI are *asinine*. "It was noted in 1858 'that the disappearance of the large quantities of game has only taken place within the last twenty years.' It was to accelerate almost beyond measure in the twenty years following."[15] "After [five] centuries of gleeful rape, the [DC] stands a mere generation [or two] away from extinguishing [millions of species of] life on this planet."[16]

The DC has completely missed the very sweetness of life, the innocence and sacredness of all creatures, the respect for the predators, the majesty of the standing trees. The DCI, its attention spent on the drive for material wealth and power, barely notices if at all. Native Americans have always known that nature literally speaks. "[A] Canadian Indian [said], 'Did you know trees talk? Well they do. They talk to each other, and they'll talk to you if you listen. Trouble is, white people don't listen. They never learned to listen to the Indians so I don't suppose they'll listen to other voices in nature.'"[17] "The Yupik Eskimos refer to... Westerners with incredulity and apprehension as 'the people who change nature.'"[18] "[Over eighty years ago], an Omaha Indian elder expressed it this way: '...but now the face of all the land is changed and sad. The living creatures are gone. I see the land desolate, and I suffer... loneliness.'"[19] This is a loneliness the DI is completely oblivious to and has no understanding of, with no comprehension of what it's doing to all life on Earth, a loneliness it may only feel when it's too late. "It is incredible arrogance towards other life that has caused destruction in this country. Who... has the right to play god and [destroy] an entire species of fellow beings... of life...?" - Jimmie Durham, Cherokee.[20]

In the 1850s, the famous Duwamish Chief, Se'atl (Seattle), whose most famous speech, interpreted and quoted many times, included that every part of the earth was sacred to his people, every pine needle, every shore, every mist in the mountains and forests. Every part of nature was revered by his people. The white man, however, was a stranger who came to take whatever he needed from the land like a thief in the night without any regard; the earth was not his brother but his enemy, and that if he continued to contaminate the environment, he would one night suffocate in his own waste. And the DC *has* continued its waste, drastically. Rather than take care of the Earth to preserve it intact for the perpetual seventh generation to come, the DC has done the exact opposite, spreading its predicted contamination everywhere. Notably, we are in the very seventh generation living since Se'atl's time.

Caroline Fraser, in her 2009 book, *Rewilding the World*, points out that: "Biologists have begun to understand that nature is a chain of dominoes…" leading to the loss of entire ecosystems.[21] So even biologists have just begun to understand what Natives have *always* known? And the rest of the DC will understand it when? After the mass extinction is over?

"It is painfully clear that the United States needs the Indians and their culture, a culture that has a deep reverence for nature, and values the simple, the authentic, and the humane." – Stewart Udall, U.S. Secretary of the Interior, 1961-69.[22] "Indigenous wisdom… extends far back into the Paleolithic Period… [and,] as the years pass… we have begun to recognize both how little we really understand these people and how much we need the wisdom of their traditions."[23] "…American society could save itself by listening to tribal people. While this would take a radical reorientation of concepts and values, it would be well worth the effort"[24]; otherwise, "Indian culture and Indian thinking, untreasured by the unknowing, will be gone, and only the history books will be able to tell us of it and make us wonder why we did not enfold it and make a place for it in our national life."[25] The Americas have had a completely ignored treasury of pre-Columbian ideology, ignored so the Treasury of Life could be stolen for the DC's self-adulation.

"Nursing anger against people long dead is a waste of one's life."[26] It's time to move on. The past is past, and since it has brought the world to an anthropogenic mass extinction, the past has proven itself to be invalid. It didn't work. The patriarchal DIC hasn't worked. The DI has had no respect for other life, thinking it was above it. To the contrary, Native American culture could be summed up in one word: respect. Respect for all life.

It's time for humankind to learn what true dominion is and start working with the laws of nature in all ways, especially for our energy needs, and use the brilliant potential of our minds to restore our Earth to health. We need to restore our population in a sane and humane way to a truly sustainable carrying capacity, which we know we're presently far beyond. We need to create north/south wildlife corridors on every continent, similar to Dave Foreman's Yellowstone to Yukon (Y2Y) vision, to replace the fractured and insufficient park systems.[27] We need to adopt and integrate many Native values.

The entire values system of the DCI needs to evolve by a quantum leap so priorities aren't led by the extravagant accumulation of material wealth while the world dies. Currently, some DC members can buy multi-million-dollar mansions and then demolish them simply in order to build newer, bigger ones, the price of which could save an entire species on the brink of extinction. The gold that some Arab sheiks use to gild their private airplanes could do the same. The $200US to buy just one bowl of shark fin soup could more impressively at least be spent on memberships to a number of environmental organizations. Also least impressive is the new proliferation in China of wildlife restaurants that serve endangered turtle and other species. They are all free to do these things as of yet, but that such desire exists is indicative of the mentally ill priorities and values of the DIC that must change.

Native Americans who have not joined or been corrupted by the DC wouldn't be using severed gorilla hands for ashtrays or purchasing skins or other body parts of endangered species to show off as macabre, so-called status symbols of the "rich." So many less drastic but common status symbols we take for granted as normal - over-sized vehicles, houses, everything needing to be new, etc. - are all passé; rather, they are signs of ignorance, selfishness, apathy, and immorality in today's world, a world of the anthropogenic mass extinction of species. A good question for everyone on Earth to ask might be, "What kind of world would we have if everyone else on Earth lived like me and my lifestyle?"

ELEPHANTS

E LEPHANTS ARE A KEYSTONE species, functioning as forest managers by knocking down trees, grazing and browsing, and eating up to 660 pounds per day, yet they only digest about 40 percent of what they eat, fertilizing the land with the other 400 pounds that are expelled. With ranges that were up to 500 square miles, these pachyderms have historically had the largest impact on maintaining and restoring several ecosystems; for instance, the famous Serengeti's celebrated abundance of life would not exist as we know it without elephants. In West Africa, one-third of native tree seeds must pass through an elephant's digestive system in order to germinate; however, people have cut down 95 percent of the forests there, and it now only has 1 percent of the continent's elephants, making the chances of native forest recovery very slim.

Elephants are intelligent, have language, are especially sensitive, exist in very intimate social units, and have a proven capacity for feeling empathy. In similarity to whales, elephants have the ability to communicate over long distances via subsonic tones below human hearing. An elephant's proboscis has 40,000 muscles, and the tip of it is ten times more sensitive than a human finger. Their feet are so sensitive they can step on an egg or dried leaves without crushing either. It takes twenty-five years before a female elephant is ready to procreate, and at twenty-two months, it has the longest gestation period of any animal on Earth; additionally, a calf is dependent upon its mother for three to four more years. Surrounded by human encroachment and hostility, this does not bode well for the regeneration of their population. In Africa in the past fifty years, the human population has grown by almost 700 percent while the elephant population has dropped by over half.

At the onset of the Golden Age of Discovery, a European global conquest of the eighteenth and nineteenth centuries, Africa and Asia had a total of 10 million elephants before man started killing them just for their ivory. Since then, the world population of elephants has dropped to 500,000, vanished by 95 percent due to humans. Africa holds over 90 percent of the world's elephant population, half of them in the countries of the South. The rest are in Asia, though now most of the young elephants there are being taken from the wild and domesticated to work for people; thus, wild elephants are a ghost species in Asia.

With a current annual decline in population of ten percent, elephants, after 26 million years of existence, could be extinct in fifty years; however, events do not occur linearly. For instance, the breakdown of government in Central Africa in the Democratic Republic of Congo (DRC) has created a catastrophic decline of 62 percent of forest elephants from 2002 to 2011, as rogue armies poach ivory to fund their efforts. This has left 95 percent of the forests of the DRC and 86 percent of the forests of Gabon empty of elephants. Together, West and Central Africa only have 7,000 elephants.[1] Also, a linear reduction does not factor in the Ghost Demand Syndrome and the Human Exponential Event. At best, elephants will most likely become a relic species such as bison in America are today.

The average weight of a pair of elephant tusks is 19.4 pounds, 56 percent less in weight than in 1970, as humans have caused smaller elephants to live and reproduce by consistently killing the ones with larger tusks. It's not uncommon for officials to find several caches per year of 35-40 tons of tusks at a time, equaling the lives of 4,000 elephants, the number illegally hunted about every six weeks. On average, one hundred elephants *per day* are killed for their ivory by poachers; that's an average of one every fifteen minutes twenty-four hours a day every day. Most of these "wild" elephants are confined to preserves, so the poachers know where to easily find them, and these preserves, known as "paper parks," are largely unprotected due to lack of funds, lack of will, and corruptive payoffs. Rather than being treated as a charismatic, flagship species with psychological importance to our species, elephants are instead wasted for the carving up of their ivory body parts primarily for Filipino religious icons (Catholic) and Chinese artifacts.[2]

Squadrons of poachers from organized crime often attack with machinegun-armed helicopters, land, cleave off the entire faces of up to thirty adult elephants (one-third of the tusk is subdermal), load up, and fly away as quickly as possible, leaving baby elephants of this extremely sentient species alive at the scene, terrified and orphaned, witnessing with all their senses their bullet-riddled, faceless parents' brains hanging out, later to often wake up screaming in captivity as a result of the trauma, literally wanting to die.

IMPAIRMENT, SUPPRESSION, DISPLACEMENT (ISD)

I SD IS THE PROGRESSION from *impairment* of satisfaction to *suppression* of use to eventual *displacement*.[1] As the world has gotten more populated and busier, those seeking quiet, solitude, and spiritual communion in nature increasingly end up with ISD. They are hikers, artists, birders, ceremonialists, foragers, and others seeking a reverent experience in nature, and what they enjoy is impaired by the presence and effects of logging, mining, fracking, four-wheelers, ATVs, ORVs, dune-buggies, dirt bikes, boaters, snowmobiles, hunters, target-shooters, mountain bikes, cattle, and people generally everywhere in nature. All these activities cause ISD, leaving ugly scars on greatly-damaged landscapes, noise, interruption, and pollution everywhere. Since many can no longer get the joyful experiences they want anymore, their use becomes suppressed, and eventually they quit going out altogether; the other activities take over to where hikers and others are displaced to the point where there's nowhere to go even if they wanted to.

The natural world has been taken over by motorization and industry, but it's not just hikers that are affected by the dangers and decimated scenery. The wildlife has already been displaced or worse. There's no natural, untouched environment anymore. Even the few wilderness areas that aren't illegally invaded are impacted. The animals have nowhere to go. The Forest Service calls all this "multiple-use."

Why are suicide rates, alcoholism, and drug abuse so high with Native Americans? They've been experiencing ISD for centuries, and it's made worse by the complete obliviousness and narcissism of the yahoo culture who have never known and never cared about the harm they've done and are still doing, who have no value for anything sacred, who just want to "tear shit up" for recreation or money. But ISD is no longer limited to Native Americans.

ARTIFICIAL INTELLIGENCE (AI)

A RTIFICIAL INTELLIGENCE (AI) IS the next big technological advance, the anthropogenic god many are awaiting to resolve all the problems we have created on Earth. AI is expected to fulfill the DC's dreams of human destiny: immortality, a new Eden, and our own extraterrestrial colonization. This ultra-intelligent super-computer is expected to be the epitome of human achievement, an autodidactic genie, and we may all be connected to it through The Singularity in the near future. Humanity will finally be liberated from our own shortcomings, and our trust can be put into knowing that all will be well, guided by our manmade, artificial super-brain.[1]

The Singularity will be the global, technological, collective intelligence and will assumedly connect anyone and everyone, enhanced astronomically with AI. Like cell phones, everyone will have access to this new necessity – the global human/AI mind - in order to survive in the new world. Many are already giddy with anticipation. AI will be the most powerful thing humans have ever created. We might even be able to create worlds of our own.

And yet, what will we make of the Earth?

Perhaps the Earth will become immaterial, irrelevant, like a cell of the sperm or ova that created us – nothing to think about, especially once we're inhabitants of the cosmos. It's an easy seduction, with expectancy of freedom beyond anything we can imagine, free of any limitations. Why bother even trying to heal ourselves and the Earth when AI will soon show us how, better than we ever could've imagined?

The question is: what values will AI have? AI will be a product of and programmed by the DI. Most likely AI will be the DIC on steroids, showing how we can do more of what we're already doing because that's what the DC wants; AI will be the DC's tool. The DC will not make AI immediately public, even though we may assume the free market or those in power will just let anyone have the power of it. But why would they? Why wouldn't the elite of the DI keep it to themselves and only slowly leak parts of AI's powers out and only if it suits them?

What if AI tells us to do all the obvious things we already know we should do but haven't because of our denial, guilt, fears, and traditions? What if AI points out our population problem once and for all, especially since everyone's going to be living forever, and tells us no one of either sex can have more than one baby in their lifetime (with forced sterilization), no matter what, until our population is in synch with our Earth's resources? What if AI also points out the futility of so many separate religions and says religion is unreasonable, too destructive, dis-unifying, and must end? In fact, AI will be the new religion, and anyone that doesn't subscribe to it will become fatally marginalized. What if it says fossil fuel extraction must stop immediately and shows how all vehicles could be run off the magnetosphere? Does the DC think the CEOs of all the polluting, mega-profiting businesses are just going to lie down and roll over?

Those of the oligarchies and the corporate plutocracy are just going to give up their power and live the lives AI has figured is the only right way to live? What if AI says that capitalism is too destructive? What if it says every adult individual on Earth must be provided a basic income for free so there's no poverty? What if it shows all money is unnecessary, since its value is supported by nothing but a normalcy-biased agreement system anyway? Governments, MNEs, and the general populace are expected to just surrender everything to AI as the new planetary monarch? More likely, the DC will relatively control AI to benefit those already in power.

Let's hope AI comes up with how to actualize cold fusion; we could feed it nuclear waste for fuel and clean up the world.

Will lifeforms other than human matter to AI? Will it care about the extinctions? Or what if it tells humanity that nature must be restored to full health and functionality, that humanity must finally learn to share the planet with other life, and that the planet wasn't made solely for us and our selfish desires and beliefs. And in order to balance the Earth systems again, that large swaths of wildlife corridors in every continent must be created free of human interference. And that the borders of countries must be readjusted and aligned according to biomes rather than human politics. What if it insists humans must give up so much of their *control*, because Earth cannot be healthy unless all of life is healthy and therefore species cannot be allowed to go extinct? Do we expect the DC to go along with all that even if it's the only smart thing to do?

It would be so much easier to preserve and restore the health of all life on Earth *now* rather than waiting for some super-intelligence to tell us we should have done it sooner. Why must we wait on an unguaranteed rescue by AI to tell us what to do? Do we really need an "outside authority" to give our power over to (and most likely argue about)? And what's going to come first, AI or the crash of the HOME? How much do we need to immerse ourselves in the tragic diminishment of the Earth and all her precious species before we voluntarily change because we already know we must?

What if AI goes the route of further reductionism, so that all we need are the raw materials for 3-D printers to create all our food or anything else we need, including new internal organs? Then we won't even need pollinators, and we can replace our livers until the anthropogenic toxins are out of our environment. Are we ready for such true science fiction?

Because of the expectation that AI is going to take good care of us and the Earth, too many think there's no reason to be concerned about global warming, radioactive waste, chemical pollution in every niche of our reality and so on. AI will simply fix everything just the way we want it…. Maybe AI will figure out how to save Neo's Ark when it's already heading over Victoria Falls…. AI will just snatch it out of the air and place it safely in a new Eden…. Even if it could or will, we will have finalized our complete separation from nature - just what the DC has wanted all along. "Hey, no problem, we'll just redefine what nature is again. After all, do we even need to be biological?" Or, as Shell Oil once advertised in *The Economist*, "Do we need nature?" (The real question is do we need Shell.)

AI will even reconstitute all the extinct species if it deemed they were useful, genetically improved of course. In fact, we don't have to worry about or change anything we're doing or become sane or responsible at all, *because...* THE AI SAVIOR IS COMING!

Big maybe.

WHALES

B LUE WHALES ARE THE largest animals to ever inhabit the Earth, larger than any dinosaur ever was. They can grow up to 100 feet in length and weigh up to *209 tons*. A blue whale's tongue alone can weigh as much as four grown giraffes, a giraffe, whose heart is as big as a sheep.

The oceans contained about 150,000 blue whales in 1850[1], but by 1990 the population of blue whales was cut to less than 400 in the Southern Hemisphere, a 99.75 percent drop. Throughout the 1930s, tens of thousands of blue whales were killed in Antarctic waters by Norwegians[2], who, along with the Japanese, have been the long-standing hunters of these oceanic mammals. As is well known, the Japanese continue to kill about 1,000 whales a year in southern oceans for so-called "scientific research." One can find their scientific research for sale on the shelves of Japanese supermarkets. This is done in spite of the International Whaling Commission's 1995 declaration of a worldwide "zero catch" moratorium on whales of any kind. Hunting blue whales was banned in 1965, which "was the broadest and most far-reaching act of kindness humanity has ever bestowed on another group of species."[3] We could do well by having such kind moratoriums on many others that are facing extinction such as Bluefin tuna and sea turtles....

All seventy-eight species of whales (eleven baleen and sixty-seven toothed) have been severely hunted. Ninety-eight to 99.7 percent of their populations were taken; some by the late 1800s, others by the late 1900s, and they are still in the process of recovering, the best at 5 percent of their former numbers. Northern Right whales, in particular, are critically endangered; Vaquita whales, which live in Mexican waters only in the north of the Gulf of California, share the same distress.

Whales reached their present, basic forms 40 million years ago, evolving from terrestrial mammals that migrated to the ocean. Similar to the case of rhinos, if 40 million years were compressed into 400 days, humans have nearly caused the extinction of whales in the equivalent of only 2.3 seconds out of thirteen months, an extremely small fraction of their ancient existence.

The U.S. military is exempt from its nation's own Endangered Species Act in the interests of national defense. This allows the Navy to carry on with a continuous practice of bombarding the oceans with the pressure of 235 decibel (Db) sound waves, which fill the oceans for thousands of miles. For five years − 2014 through 2018 − the Navy will be conducting about 3 million of these blasts, averaging over one per minute. This is pure torture to whales, dolphins, and much other sea life, including invertebrates. Such sound blasts disrupt their diving habits and have given whales and dolphins symptoms of the bends, which causes their brains to bleed and bubbles in their blood to wreak havoc on their internal cells. Decibels of 210 can cause their lungs to rupture. These mammals' lives depend upon their ability to hear, but the U.S. government is allowing its Navy to cause whales and dolphins to go deaf and die.

Whales have sung their underwater songs uninterrupted for tens of millions of years, but now, due to anthropogenic events in the ocean, thousands of these magnificent, wild, intelligent mammals have despondently ceased to sing at all.

SOLUTIONS

"Until one has loved an animal, a part of one's soul
remains unawakened." - Anatole France.[1]

"If anger… [isn't] appropriate at times like this, then we humans
are truly worthless, and we may as well curl up and die."[2]

"If you know what the most important thing in the
world is that you could be working on, and you're not,
why aren't you?" (Paraphrased) Aaron Swartz[3]

WHAT KIND OF WORLD do we want?[4] We live in a time when the dominant culture is in an ever-increasing race to the bottom. The bottom, where only weed species will exist in destroyed ecosystems, all. Yet that very culture is what most people blindly put their trust in, take their orders from, and form their belief structures by without questioning anything, as if they had nothing to do with why the human-caused mass extinction of species is happening. To continue to go along with the status quo is to be part of the problem.

If we do nothing different, the world continues its path of the HOME by default, the sixth mass extinction on Earth *in our lifetime*. We are already immersed in its well-matured beginnings. We know what it looks like and what it's growing up to become. "[W]e don't have to do anything to bring about this future. All we have to do is nothing. Just continue to do what we are doing now, whether it's counting on a techno-fix or tending to our gardens or telling ourselves we're unfortunately too busy to deal with it."[5]

Our time on Earth gives each of us now an unavoidable task. "[A]ll citizens who have moved beyond denial... understand that current dominant practices by humans have put us on a collision course with Nature."[6] "The success or failure of any historical age is the extent to which those living at that time have fulfilled the special role that history has imposed upon them...."[7] The calling of our times is much bigger than our own individual lives, yet our own repercussions, value, and meaning depends upon how we answer.

"A time comes in the history of souls, as of nations, when forbearance ceases to be a virtue, and self-respecting life is only to be retained through defiance of and rebellion against, existing customs."[8] "A time comes when silence is betrayal." - Martin Luther King, Jr.[9]

Deep down, we know the Earth is being decimated, but most are choosing not to pay attention; it's too painful or too *inconvenient*, and even most that do pay attention aren't doing enough. It is not enough just to know about the HOME; healing requires action to take place. If you're sick, you don't just admit it, you do something about it. The truth alone will not set you free; truth requires action to be taken if we want a healthy world. As Tim DeChristopher (the heroic protester that successfully disrupted and exposed the illegal, governmental auction of public wilderness lands in Utah to energy-company bidders) said in 2008, "At this point of unimaginable threats on the horizon, this is what hope looks like. In these times of a morally bankrupt government that has sold out its principles, this is what patriotism looks like. With countless lives on the line, this is what love looks like."[10]

It was Shakespeare that wrote, "When law can do no right, Let it be lawful that law bar no wrong." More than crimes against humanity or crimes against nature, both of which harm all life on Earth, the greatest crime is against Creation itself by causing and allowing entire species to go extinct, especially when we consciously know we're doing it. When does action against such crime supersede human law? We must at least make it a felony to willfully cause species to go extinct. It's already a capital crime in China to kill pandas. And if lawmakers never enact such laws, what then? Far too much is irretrievably at stake to do nothing, and at what point are we also the immoral ones to comply with the human-caused mass extinction of species? After all ecosystems are ruined? After 50 percent of primates are extinct? After the Amazon rainforest is gone? After tigers, leopards, and jaguars, and even giraffes are extinct in the wild?

All the change that needs to happen is up to the people. Nothing less than a mass movement – a "spirit of the times," the planetary awakening of humanity, if you will – is all that will cause successful healing: "…an awakening of the human consciousness to humanity's deeply destructive path, the shallowness of lives devoted to mindless material consumption, and… creating a world that values life more than money."[11] Our priorities are all askew. "'Use' as our primary relationship with the planet must be abandoned"[12]; such is an ideological shift. It's not ok to just take forever or to waste massive amounts of money on militaries, sports stadiums, recreational drugs, endless developments, and on "status" consumerism, all of which could rather be spent in a plethora of ways to heal our HOME.

The question is: What are you doing to heal the Earth and avert the HOME? There is nothing else left to do. We are fast approaching the apex of the crisis. From there it's downhill because of the debts we've caused: the environmental debt, extinction debt, carbon debt, and so on. It's like smoking finally catching up with the smoker; at a certain point, though they still appear to be "normal," they've got undetected (or denied) stage-4 cancer and are gone in weeks.

The status quo is a death trap, a dissolution to the lowest common denominator where complete orders, classes, and families of life will be lopped from the genetic tree; the limbs and branches are already beginning to fall, and the tree is being poisoned from its roots. To ignore the duty upon us, the duty which is ours *now*, is to forsake not only your own life, but your children's, your grandchildren's, and it is to forsake the species of the world – the amphibians, pollinators, caterpillars, trees, penguins, coral, beetles, the hundred in *Going, Going, Gone?*[13] and the *tens of thousands* of others already vulnerable or critically endangered and the *millions* more approaching the same fate.

The functional motto has always been: "If it's not good for the Earth don't do it," though too many have not wanted to acknowledge and admit it (overpopulation, for example, could explicate the faulty dynamic of the entire DI), and most of the world doesn't even have a choice. We have many problems we can't avoid and can no longer afford to deny. Those days cannot go on. Doing as much as possible to heal our Earth is our priority. Preserving our Earth was never the priority of the DC, which has gotten us into the extinction debt we're in, and we *must* correct the damage as soon and as much as possible.

The good news is the Earth will mend itself amazingly if we give it the opportunity. Our Earth can no longer afford anything less. When you become part of the movement to heal our Earth *which must occur*, you will feel the deepest and most fulfilling purpose to your life and be part of history-saving creation rather than part of the old DIC. Every one of you has your role like cells in a body, and each cell must be healthy, too, for the whole. We must get serious now; the inevitable is here… with billions of lives on the line.

As inconvenient as it is for most – especially the comfortable – you must come out of denial and quit believing in the narcotizing corporate commentators, the cultural norms, the abase consumerism, and incognizant leaders of all kinds. You are living in a time where you must educate yourself to what's going on: "We cannot care deeply if we… have no knowledge of that which we are asked to save, and if we do not care, we shall not help."[14] If we don't help – and help *now* - our once-paradisiacal Earth is destined for complete ruination, such ruination which could take millions of years for the evolution of nature to restore Earth's healthy balance.

All you've really been told to do is to change your light bulbs, turn down the thermostat in your home, recycle, drive less, and use less water. Big deal… plus most of you don't even do that much. That's like saying, "It's easy to be healthy – smoke one less cigarette a month."[15] It's as if our Earth has a malignant brain tumor (the DC ideology), and we're plucking our eyebrows to fix it in "selfie" cultures that are becoming more narcissistic every day. These band aid habits we're told to do achieve relatively nothing. For one, using less electricity doesn't save any. A certain oversupply of electricity has to be generated to guarantee that demand is met, but all the electricity not used is not saved anywhere; there are no storage batteries at the end of the line. Cutting back on your electrical use is good, but it only allows for the increasing demands of a growing population to join an unchanging system. More power plants will still need to be built. (If you really want to save electricity, get rid of your TV! Your TV uses far more electricity than your lightbulbs. It will also get rid of the consumerist, political, brain-numbing propaganda pouring into your home, improve your mind immensely, end a serious addiction, and free up your time to do something worthwhile that life on Earth needs, including yours.)

It's up to the people to create a mass movement because the political system is reckless, corrupt, and self-serving; its evolution now keeps the public ignorant, confused, and indifferent, and it has reduced the populace to irrelevancy. "The democratic system... that once made piecemeal reform possible... exists only in name. It is no longer a viable mechanism for change."[16] Priorities are aligned with a very destructive ideology which ignores the state of life on Earth. "Destroying life to make money is a social pathology."[17] Politicians are compromised and unwilling to change, as unwilling as the corporations that control them, and neither are going to change except to "greenwash" their same old activities so they can continue doing them. Under the current system, to change would marginalize MNEs out of their industries, and to change enough would end their current operations.

And yet we cannot keep doing what we're doing.

The carbon exchange is a great example of greenwashing to preserve the current system in perpetuity. The DC and its environmental organizations even think they're very clever in having dreamt it up. A business can offset its over-pollution by buying credits from another business that is polluting under its limits, and it can also get credit off its pollution by declaring any forests it may own for absorbing carbon just by growing. In other words, healthy forests now count as polluters, via pollution credits, that make money for polluting industries to get around environmental laws. Carbon pollution was even traded on its own stock market, the Chicago Carbon Exchange (CCX) for investors to reap large sums of money. The CCX traded $21 million on its first day.[18] Brilliant... a new, lucrative market for polluting was invented to circumvent any positive change. This is just one example to point out that so far, nothing – *nothing* - has been done to preclude the worst of the foreboding HOME. Even those organizations that calculate your "carbon footprint" don't even mention factoring in how many children you've had, which is by *far* the largest footprint you could make.

Additionally, most of the large, long-standing environmental organizations, though somewhat successful, are now fully within the DC system, having been infiltrated and funded by the corporations they are meant to regulate, and the fat cats in charge are personally benefitting profoundly in the name of conservation like any other globalized MNE.[19] Better than nothing, environmentalism has become big business.

Those running the dominant system are just trying to buy time because they're unwilling or unknowing of what to do; as long as they can convince you that everything's fine and keep you in denial, they can keep making their profits, living the good life, and further their agenda.

Again, only a people's movement is going to make things change. "Saving biodiversity will have to become a global obsession, not merely a pastime."[20] "[N]o government undertakes a question of a really important and social character until it has been forced upon it by the voice of public opinion."[21]

You can easily take the first steps to become involved. There are 12,000 environmental groups just in the U.S. all with very inexpensive membership dues, if any. *Consider it mandatory to join several.* PAY ATTENTION! There's so many to join, even the big ones are better than nothing, and at least you'll begin to become informed and involved. Also, even the largest environmental organizations have less than 1.5 million members; that's less than half of one percent of the population. Think of the impact those organizations could have if that number rose to ten percent or more. Imagine 50 percent. It's a very small but worthy, caring duty to give back to your Earth. Any other options to choose not to help heal are reckless.

Recognize that people of the lowest status are those who are still doing what is harmfully obsolete and doing nothing to heal our HOME, or, in fact, are continuing to perpetuate its demise. Just one example (of millions) is one of the low-status members of wealth of the Dallas Safari Club buying a $350,000 permit from Namibia to hunt down an endangered black rhino "for conservation purposes."[22] It goes on and on.

It's up to us to create demand for healing solutions and be intolerant of driving gasoline cars that only get seven to eighteen miles per gallon. Is such apathy and ignorance *really* the best we can do? The questions at social events need to be less of, "What do you do for a living?" and more about, "What are you doing to help heal our Earth? How many worthy environmental organizations are you a member of? What's the best, most effective letter you ever wrote to a Board of Directors? Do you have solar panels on your house? Must you build yet *another new house*? Do you drive an electric car? Have you gotten rid of your lawn yet and replaced it with native species? What great books have you read on healing our Earth? What are you doing about the HOME? Have you pulled your financial investments out of fossil fuel MNEs? What are you doing to inform yourself and others of what's really going on? What are you doing or think about our overpopulation? Do you still support nuclear energy and actually believe it's a clean energy like you've been told? Have you ever heard of Tesla? Do you ever even question your culture's values?" Get used to planting such seeds, or get used to the status quo ruining your life with weeds.

It's also ironic. All those Native Americans you've either denigrated or wished you could be in your imagined fantasies are the ones who have the greatest potential to stop many of the DC's plans because the Indians have protective laws on their side, at least on paper. All you who have long thought that none of those Indian treaties should apply because they were made so long ago might want to realize that you're still profiting from that 1872 mining law and the 1886 Supreme Court ruling that made the Bill of Rights applicable to corporations as persons. Native Americans may be our best legal hope, and they have always been our greatest ideological hope, you just haven't listened. They are also the ones in the greatest need of finances in order to fight the DC in court. Not only do they deserve to have your support, the well-being of our planet is dependent upon it. All they're doing to preserve their very lives is what will help us all.[23]

We all need to demand that the Tesla energy patents come to light. You've been led to believe he was a crackpot, but meanwhile, you're using his inventions of electricity, x-ray, radio (although Marconi got credit), and up to 1,400 more in your daily lives. His greatest inventions are still suppressed, which would make even "alternative energies" an extraneous point. If you don't believe any of this, *inform yourselves*. Other than suppression of technology and greed, there are no reasons for using fossil fuels for our main source of energy, *which is destroying our planet*.

All heating oil should be replaced with 100 percent biodiesel fuel (if not solar). This could easily be done.

Find and buy your gasoline from gas stations that do not sell ethanol-added gasoline.

Building codes should make solar power mandatory, require systems that combine a secondary grey water system, and make flush toilets obsolete, replacing them with high-tech commodes that could even generate energy. These new codes need to be applied to every new and existing home. That would put millions back to work.

Ships must be required by law to be refitted to process their own garbage and sewage to keep it all from being dumped into the ocean.

Stop using herbicides and pesticides on your lawn or property! Boycott all such chemicals. You think it's normal to use them, but they're killing lives on Earth, both human and non-human. They're even killing you. Quit buying poison!

And the OFCs ought not try to copy or emulate the OCCs. Much as they have done with communications, having gone from A to Z without experiencing the middle stages, the OFCs could leap-frog over the OCCs and lead them into doing things differently. And it is essential; for instance, the world can't afford to have a billion more flush toilets or gasoline vehicles or more coal or nuclear fission plants or paper made from trees.

Militaries from around the world are going to have to be redeployed to start protecting the lives of endangered species from illegal human activities. "...Stewart Brand, the former military officer who gave us the original *Whole Earth Catalog* [said]: 'My platoon could have made short work of restoring a salmon stream, assisting a controlled forest burn, helping protect African wildlife from poachers, or planting native shrubs at the edge of the growing desert. I wonder if they might get this opportunity.'"[24] Are we going to let rhinos and gorillas, for instance, just go extinct? What's more important, our self-absorbed focus on wars and human drama or the preservation of species from permanent annihilation?

We could create an international Green Cross to restore habitats and their native species and create wildlife corridors around the world, "at least until our armies expand their role." How many citizens and professional engineers and others "would truly love to... restore rivers... [and] use their skills to rip out unnecessary dams," and restore salmon runs, for example?[25] Even after over seven decades since the Grand Coulee Dam was built without fish ladders, as well as too many other dams around the world, the DC still hasn't found it important enough to build fish ladders around them. This is just absolute stupidity. The DC would rather allow salmon to go extinct! Such dams without fish ladders should be a crime. Mistakes of such magnitude *must be corrected*. Either take the dams out or fix the problem that's causing such extinction and disruption to the natural world. Salmon are another keystone species! The DC's priorities of putting nature at the bottom of the list are a mindboggling and pervasive cultural mental illness.

Religions, which were created before science even made its most rudimentary discoveries, must be brought up to date. An excellent example of this is the great work of Christian pastor Michael Dowd. His personal creed is, "the work of ensuring a just, healthy, beautiful, and life-giving world for future generations of all species."[26,27] His work is brilliant and much needed.

"Boycott [all] products that contain palm oil."[28] The creation of palm oil plantations are destroying rainforest habitat that are home to critically endangered orangutans, including many other species. It's probably even in your shampoo (sodium lauryl and/or laureth sulphate) and peanut and almond butters.

The magnificent ancient temperate rainforests of the Tongass of Alaska and the Primorye of Russia, along with the similar, smaller forests of Chile and Argentina, need to be immediately preserved in any way possible. The amount of these forests combined add up to only 3 percent of what's left of tropical rainforests; are of indescribable, unparalleled beauty; and are truly important heritage sites, sacred sites, natural "cathedrals" that far surpass any cathedral man has ever built, and they exist *nowhere* else. They also contain many critically endangered species and centinelan species. It is tragic senselessness - even criminal to the world and the future - that these priceless forests are as yet allowed to be commercially and illegally exploited.

Convert over to an electric vehicle now and demand that solar panels be incorporated into them. Demand the development of solar paneled roads. Such panels may already be available, and if they're not yet viable, it's a great technology for the government to advance. New advances in batteries for solar energy are now available that can revolutionize how we produce all our energy, including powering our vehicles.[29] Many have been led to believe that electric vehicles are no better than gasoline or diesel powered ones, but fossil-fueled vehicles still burn as inefficiently as they did 100 years ago (less than 20 percent of the fuel is used to power a vehicle while the rest is wasted, non-recyclable, and lost forever as pollutants into the environment); meanwhile, batteries are recyclable and waste no energy and spew no toxins when handled right. Plus, higher demand for electric vehicles would greatly help change our energy paradigm.

Pull all your stock investments out of the fossil fuel companies, nuclear plants, weapons manufacturers, coal mines, etc. Write to the Board of Directors explaining why you're doing it, and tell other companies you stay invested with what you want them to change, or you're out.

Write your Congressional representatives; if they hear from enough of you, they'll finally start to listen. The people must start a wave or nothing's going to change, even if it takes millions in the streets. Millions of people are already being displaced by the way things are. This is not to say "the end is near"; it is to say that we're going down a very mistaken, tragic path.

Demand that the wrongful and ludicrous law upheld by the Supreme Court of 1886 must be changed so that MNEs are not considered as persons with the same individual rights as people. *Citizens United* and *McCutcheon* also must be overturned.

It's also no longer apropos, responsible, nor even moral for anyone – especially the highest religious leaders - to condemn the use of birth control. Lack of birth control is the primary cause of the HOME.

OCCs need to commit $100 billion per year to the OFCs for

economic aid to preserve their environments and non-human species. The Vatican and its bank, for instance, could become wildly popular even by its detractors if it came clean and agreed to purchase the rest of the Amazon with part of its secreted wealth, preserving and protecting the tropical forest and its species – God's creation - from extinction indefinitely. Maybe it wouldn't even have to do it alone. What if the Vatican started a movement where 1.2 billion Catholics averaged contributing $10 a year to buy the remaining Amazon? What if we all contributed? Why are the governments and people of the world not already doing so? All those species going extinct are species of the world – of Creation – not just species of individual manmade countries.

Imagine how incredulous people of the near future (if not now) will be when they hear about the extravagant use of water solely for golf courses. We need to stop creating new golf courses and return existing ones into parks or preserves with native species instead of wasting so much water and land, especially in desert biomes. This is where eminent domain would make sense. Up to a million gallons of water per day are needed for golf courses. Critics will say, "Yeah, but now the new ones are required to use treated water, so it's ok." That treated water gets used once then soaks into the ground. And of those million gallons, how much is treated? Ten percent? With whatever percent, it can then legally all be called treated. A million gallons still needs to be used every day. There are so many better uses of land and water! An average of at least twenty-one species goes extinct during your round of golf. Is that *really* your best use of time and resources in light of our HOME's situation?

Again, stop eating beef. Second to humans, cattle are the most destructive species of the planet. Also, if we didn't raise so many cattle, up to 50 percent of fresh water use would be available for other purposes or saved. Refraining from beef could also dramatically improve your health.[30]

Replace all the corn that's being grown for ethanol with hemp. Hemp generates four times more pulp to make paper with than trees do and is not of the quality for drug use. We could save forests by growing vast amounts of hemp, get rid of the highly poisonous corn-ethanol industry - which is also all GMO corn and inedible unless highly processed into something else - and return soils back to being organic in a few years for other healthier crops.

We must create north/south, continental, contiguous swaths of land - wildlife corridors where people are not allowed. Only the truly indigenous people that are left could remain, but we would have to support them by appropriate means so they do not kill endangered species; they, too, may have to change some ancient customs. We cannot allow species to go extinct. The DC has never realized Earth has to be shared with other species besides its domesticated own. It does no good to reconstitute extinct species from vials of DNA (success with that being a long shot at best) if there is nowhere to put them where they can thrive and restore ecological balance.

Further development of the natural world must stop, especially in the case of wild wetlands. We must entirely cease draining and developing wetlands. Wetlands already lost around the world equals the size of Canada, enough to change the climate and the natural world itself. We have and continue to oversimplify the complex, interconnected processes of nature to great detriment, and it is already not working, extinguishing an untold amount of species and the health of our planet.

It took a massive effort to win WWII, and that's the type of effort we're confronted with now. We've created gigantic, life-threatening imbalances to our planet. We must do whatever it takes to regain a healthy balance for the nature of our Earth. We can't fix our problems with the same mindset that created them; a paradigm shift must take place. Our current paradigm has utterly failed. Humanity must realize our human-caused ongoing mass extinction, our true plight, and truly unite as never before.

If ETs were causing a mass extinction of species for whatever reasons they had, we might actually all unite against them over it, but rather, we're doing the same thing to ourselves. We're on a suicide mission and don't even realize it, taking away the very beauty and functionality of our planet. MNEs characterize such figurative ETs, and they don't need national or political borders. The concept of private property rights is not working, either, especially in regard to MNEs; in no way should such "rights" justify criminal activity to destroy lives on Earth, both human and non-human. Such "rights" must be changed. "Political rights belong to people, not to artificial legal entities. The claim by corporations to the same constitutional rights as natural born persons is a legal perversion without moral or logical foundation. ...[C]orporations should obey the laws decided by the citizenry, not write those laws."[31] And as long as corporations are legally regarded as persons, they should be tried for murder, which, as for persons, has no statute of limitations.

Rearranging political borders that are based upon and around biomes need to happen at some point, the sooner the better. For instance, "...proximal Canadians and Americans identify better with each other than with their own countrymen living farther away."[32] Southern Oregon to the Alaskan panhandle and British Columbia could be a new political border, ending at the crest of the Cascade mountain range; no change in language even needed. It could be called the state of Cascadia. People of the rural Eastern Washington and Oregon plains and desert would love it, since they always lose every election because their voters are vastly outnumbered by the urbanized people of the temperate rainforest of the western half of the states. It's an environmental problem because the land and the issues are completely different. The point is we need new ecological states, not old and dysfunctional political states bordered with straight, squared lines dreamt up by the minds of men that had nothing to do with nature. An ecological dead zone by any political name is still a dead zone. We have overlaid an anthropogenic mental grid, disconnected from nature, over nature's naturally functioning boundaries. How about the entire Colorado River drainage system becoming one or two (upper and lower) states? How about the entire Sonoran Desert becoming one state so some giant wall isn't needed that kills endangered species.... Similar border restructuring could be done around the world. Remember, "[w]e are truly on the brink of a catastrophe. In such a situation, radical change... is the only hope for survival."[33] "Shouldn't we consider in every nation major changes in the traditional ways of doing things, of fundamental restructuring of economic, political, social, and religious institutions? ...Fundamental changes in society are sometimes labeled impractical or contrary to human nature, as if nuclear war were practical, or as if there were only one human nature. ...Our loyalties [must be] to the species and the planet."[34] – Carl Sagan

There are countless solutions to all our problems, much more than any one person can imagine; that's why *you* are needed. All of you are needed. Most of all, our work is to deeply change the current ideological paradigm. Everything needs to be reevaluated. "We cannot solve our problems with the same thinking we used when we created them." – A. Einstein. We really can't let anymore species go extinct! It's bad enough already, and half the species we have now are very prone to permanently disappear within about sixty years.

Women, and the reawakening of the true feminine within all people (what Carl Jung called the anima, which we all have), may be our greatest hope for changing the DC's ways of thinking. It's patriarchal thinking which has given us the HOME. It is the heart of the feminine which holds the greatest strength and courage of the world when the lives of those they love are threatened, and it is by no means weak; it is the patriarchy which has defined and judged it as such and has suppressed feminine power at all costs for millenniums. This power must come forward now and make a stand, a Stand of the Ages, a stand to create a true paradigm shift to prevent what is already set in motion, a stand to create the loudest universal "NO!" ever heard and enacted. What is going on is *not acceptable*. Failure is not an acceptable option, a failure which destroys half the species on Earth, and a failure in the short-term which destroys the well-being of our own and certainly our children's and grandchildren's lives.

Again, *"what kind of world do we want?"* Most of all, we must create a new ideology. "Technology is not going to save us. Our computers, our tools, our machines are not enough…." Joseph Campbell.[35] We must bring a new world into being. Imagine news headlines in the near future, headlines we would call miracles, miracles which drastically restore our hope and motivation for an actual new world: First Continental Wildlife Corridor Designated In North America, Other Countries To Follow; Vatican Leads World To Purchase Amazon Rainforest Protected Intact For All Time; Indonesia Concedes Orangutans More Valuable Than Palm Oil; Smart Phone Currencies End Power Of Central Banks (already beginning in Kenya); First Magneto Vehicles Hit The Road, All Others Obsolete; Leaders Of All Religions Officially Approve Use Of Birth Control; Magnetosphere Supplies All Power, International

Agreement Closes World's Nuclear Plants; Alaska's Tongass Receives Permanent Protection Under World Heritage Status; Pope Defines True Meaning Of Dominion, Corrects Fundamental Ideology.

If these ideas sound far-fetched, it indicates how normalized and numb we have become, for they are the very type of things that must happen; however, we must learn about, demand, and create them. Otherwise, must we buy into using fossil fuels until they are exhausted only so corporate higher-ups won't be deprived of their obscene and criminal lifestyles? Do we want to be enemies of other countries forever? Do we want to continue to spend more on militaries than any other expense for all time? Do we want to have a world where half the former species of this gorgeous Earth are dead and gone forever? Must we take instructions from ignorant and corrupt religious, political, and corporate leaders who know less than we do and mislead us on purpose for their own self-serving, fear-based, separatist, and elitist agendas until the Earth's demise? Must we continue to believe that having as many children as possible is a sacred duty (according to who?) and the measure of personal pride and cultural success? Must we take "legal" pharmaceutical drugs that have never been properly tested because we're told they help us cope with (their) reality and become even more powerless? Must we unquestionably put up with cultural lies about everything even unto our death? Must we tolerate half the human population in poverty when the entire world monetary system is based on nothing but a literal illusion? And since it is an illusion, why don't we change it to something that actually works? Must some of us insist on having thirty pairs of footwear and twenty different purses in our closets and bigger houses and bigger cars and brand new everything at all times? Must we sit powerlessly in front of our TV screens and watch the well-being of our Earth slip away, species after species, disaster after disaster, seeing some occasional show the DC puts on to give us a false hope that everything's ok and that someone *else* is at least doing something good for the world? Must we define success as how "wealthy" we can show ourselves to be (as defined by the depraved DIC) rather than how generous we are or how connected we are to The Great Mystery/God manifest in all life and beyond?

Otherwise, our collective vacuity, our apathy – apathy being the opposite of love - will be exemplified by the subsequent vacuity of our HOME, ultimately revealing our humanity's collective spiritual degeneracy. Spirituality: what the DC ironically and arrogantly heaps pride upon itself for being our species' specialty and unique gift from God.

LIST OF ABBREVIATIONS

BP: Before Present, rather than AD and BC.

The following (4) are synonymous and used interchangeably:
    DC: Dominant Culture
    DCI: Dominant Cultural Ideology
    DI: Dominant Ideology
    DIC: Dominant Ideology of the Culture

GDS: Ghost Demand Syndrome, when an endangered species becomes a ghost species its value goes up to the point that it becomes more sought after before it becomes extinct, further ensuring and quickening its extinction.

GY: Galactic Year, or 240 million years, the time it takes our solar system to orbit our galactic center.

HEE: Human Exponential Event, the exponential growth and activities of the human population and its omnipresent, planetary results.

HOME: Human-caused Ongoing Mass Extinction

IUCN: International Union for the Conservation of Nature

MNE: Multi-National Entity, a globalized corporation.

OCC: Over-Consumerized Country, rather than a developed country.

OFC: Over-Fertilized Country (Overpopulated), rather than an "undeveloped" country.

APPENDIX, SPECIES TYPES

Critically Endangered Species (simplified) are those who've had a 90 percent reduction in population in the past ten years; have a range of less than 100 square kilometers (thirty-nine square miles or 6.2 by 6.2 miles), which may be severely fragmented, or occupy only one area on the entire Earth of no more than ten square kilometers (3.9 square miles or 1.98 by 1.98 miles); and have fewer than 250 adults, with no more than fifty adults in a subpopulation. Their chances of going extinct in the wild are at least 50 percent in ten years or three generations, whichever is longer. (Two lists[1,2])

Endangered Species are those who've had a 70 percent reduction in population in the past ten years; have a range of less than 5,000 square kilometers (1,950 square miles or 44.2 by 44.2 miles), which may be severely fragmented, or occupy no more than five areas of no more than 500 square kilometers each (195 square miles or fourteen by fourteen miles); and have fewer than 2,500 adults, with no more than 250 adults in a subpopulation. Their chances of going extinct in the wild are at least 20 percent in twenty years or five generations, whichever is longer.

The list of endangered species in the U.S. is currently 1,361, with 286 candidates; however, the number would be at least four times that (up to 6,500) if it were not for the economic concerns of special interests and politicians, but the citizenry is uninformed and without the will to demand species' protection. Of the already compromised endangered species list, the entire U.S. Wildlife Refuge System can only help maybe one out of five of the listed animal species. There have been only thirty-seven endangered species successfully removed from the list in forty-two years; we're obviously falling far behind. "[T]he Convention on the

International Trade of Endangered Species (CITES)… [lists] about 30,000 plants and animals… [and] thousands of species are added annually."[3]

Vulnerable Species are those who've had a 50 percent reduction in population in the past ten years; have a range of less than 20,000 square kilometers (7,800 square miles or 88.3 by 88.3 miles), which may be severely fragmented, or occupy no more than five areas within a limit of 2,000 square kilometers total (780 square miles or twenty-eight by twenty-eight miles); and have fewer than 10,000 adults and no more than 1,000 adults in a subpopulation. Their chances of going extinct in the wild are at least 10 percent in 100 years.

Threatened Species include those which are vulnerable, endangered, or critically endangered. The Endangered Species Act (ESA) considers species threatened if they're likely to become endangered without intervention. The status of vulnerable species alone is already tremendously bleak; the adult members of a vulnerable species would only fill 15 percent of the seats in only one professional football stadium (they're outnumbered by humans 730,000 to one). Endangered species have it four times worse than vulnerable species (outnumbered almost 3 million to one). Critically endangered species have it fifty times worse in area and ten times worse in adult population than endangered species; they are 200 times worse off in area and forty times worse off in adult population than vulnerable species (outnumbered over 29 million to one by humans).

Benthic Species refer to those that live on or near the bottom of the ocean. Examples are sea urchins, starfish, cod, halibut, coral, sponges, and untold more, including centinelans.

Charismatic Species, also known as charismatic megafauna, are species such as panda bears, orangutans, jaguars, elephants, gorillas, tigers and others that we tend to identify with anthropomorphically and aesthetically, hence valuing them over other species, often making flagship species out of them.

Crepuscular Species are active during the terminator, the hours between night and day of dawn and dusk.

Diurnal Species are active in the daytime, such as butterflies.

Endemic Species are unique lifeforms that are native and exist in only one particular area of Earth, oceanic or terrestrial, such as a particular valley, cave, mountainside, lake, coral reef, or island, etc.

Exotic Species are another name for invasive species.

Flagship Species are wild animals that have a type of celebrity status such as spotted owls, wolves, tigers, bighorn sheep, and polar bears. They become public symbols that gain enough support to influence political change in favor of conservation efforts, which benefit all species within specified habitats.

Foundation Species is another name for keystone species because they are the foundation of an ecosystem.

Freshwater Species have more species in risk of extinction than in any other type of ecosystem.

Ghost Species are those whose extinction is certain for a variety of reasons. There are not enough conservation programs for all the relic species, and such programs are not always successful. This is where the extinction debt applies. Some species may appear healthy now but their regenerative capacities have been destroyed. Once their present generation dies off, there are no younger generations to replace them, or their populations comprise too small of a gene pool, or they live in too small of isolated islands of habitat. There are many flora that live for decades that fit this category. Within our lifetimes, half our Earth's species are expected to become ghost or relic species or worse.

Indicator Species represent the status and future of an ecosystem and its other inhabitant species. An example is the Northern Spotted Owl in the northwestern U.S. whose decimation of population showed how much the entire ecosystem had been devastated by rampant private, state, and federal industrial logging.

Introduced Species are only slightly different from exotic or invasive species in that they are intentionally released into a new habitat from somewhere else. One example is the mongooses that were introduced to Hawaii in an attempt to control the rat population, an idea which completely failed and made things worse, as one species is generally diurnal while the other is generally nocturnal, though both could be considered weed species or pests.

Invasive Species are species of all kinds that find a footing and take over in new territories after the destruction of habitat by human activity. They often include weed and introduced species and are the second-leading cause of the extinction of native species. Global in scale, they "prey on the native animals or lay waste to the plant life… caus[ing] an estimated $1.4 trillion in environmental and economic damage each year."[4]

Linked Species are those which have specific symbiotic relationships; their existence depends upon specialist species. Examples are the *Tegeticula* moths: *T. maculata* pollinates only the *Yucca whipplei* plant, which in turn supplies only that moth with its seeds used for the moths' reproduction. Forty different yucca species depend only on the genus of *Tegeticula* for pollination.

Nocturnal Species are active in the nighttime, such as moths, which outnumber butterflies ten to one.

Pelagic Species are the ocean creatures that travel the high seas, the open oceans, such as Bluefin tuna and whales.

Pollinator Species include many species, such as moths, butterflies, bats, hummingbirds, lemurs, beetles, bees, and other insects.

Relic Species are considered genetic dead ends and include some charismatic megafauna and hundreds of thousands of other species whose populations are decreasing. They live marginal existences in small islands of the "wild," in zoos, preserves, shelters, or in permanent and direct management that requires captive breeding. They would become extinct without any of this help. The California condor and Javan rhinos are classic examples, but relics also include thousands of plants. Often zoos and other such management programs are relic species' last phase before winking out forever.

Specialist Species are flora and fauna that require a very particular environment in order to exist or have a very specialized niche or unique service to provide. Examples are freshwater cave-dwellers; pollinators that service just one species of plant; or any of thousands of other exclusive, symbiotic relationships.

The Living Dead are species that are not capable of surviving on their own because their food sources are already extirpated or extinct; they are expected to go extinct because they are wholly dependent upon human intervention, including captive breeding.

Umbrella Species are similar to foundation species, but rather than creating environments, their functions maintain and keep their ecosystems healthy, which in turn benefits innumerable other species.

Weed Species are those which thrive off human environmental activities and might be considered pests. They include invasive species of plants, insects, mammals, birds, fish, mollusks, and other organisms that follow in the wake of globalism, habitat destruction, urbanization, and more. Weed species are generalists that opportunistically fill in the vacancies where native species start to disappear or are crowded out. Examples include cheatgrass, rats, crows, kudzu, pigeons, coyotes, cowbirds, raccoons, starlings, spotted knapweed, feral domestic animals, and many others. Their prevalence is reorganizing the web of life to a globally homogenized status of compromised biodiversity and health, in a sense, developing a lowest common denominator.

Homo sapiens are just one species out of an estimated 9 million, including centinelans, yet we are on our way to carelessly annihilating half of them, day by day, in just a few decades.

NOTES

URGENCY, PART I

1 Dudley, William (Editor). *Biodiversity*. San Diego, CA: Greenhouse Press, 2002.
2 Corwin, Jeff. *100 Heartbeats: The Race to Save Earth's Most Endangered Species*. New York, New York: Rodale, 2009.
3 Berry, Thomas. *The Great Work: Our Way into the Future*. New York: Bell Tower, 1999.
4 Butler, Tom. *Wild Earth: Wild Ideas for a World Out Of Balance*. Minneapolis, MN: Milkweed Editions, 2002.
5 Lambin, Eric. *The Middle Path: Avoiding Environmental Catastrophe*. Chicago University Press, 2007.

AMPHIBIANS

1 Chivian, Eric and Aaron Bernstein. *Sustaining Life: How Human Health Depends On Biodiversity*. Oxford, New York: Oxford University Press, 2008.
2 Kolbert, Elizabeth. *The Sixth Extinction: An Unnatural History*. New York: Henry Holt and Company, 2014.

CHEMICALS

1 Corwin, Jeff. *100 Heartbeats: The Race to Save Earth's Most Endangered Species*. New York, New York: Rodale, 2009.

[2]   Ayres, Ed. *God's Last Offer: Negotiating for a Sustainable Future*. New York, NY: Four Walls Eight Windows, 1999.

URGENCY, PART II

[1]   Devall, Bill. *Clearcut: The Tragedy of Industrial Forestry*. San Francisco: SierraClub Books/Earth Island Press, 1994.
[2]   Mason, Jim. *An Unnatural Order: Why We Are Destroying the Planet and Each Other*. New York: Continuum, 1997.
[3]   Smith, Laurence. *The World in 2050: Four Forces Shaping Civilization's Northern Future*. New York, N.Y.: Plume, 2011.
[4]   Corwin, Jeff. *100 Heartbeats: The Race to Save Earth's Most Endangered Species*. New York, New York: Rodale, 2009.
[5]   Foreman, Dave. *Rewilding North America: A Vision for Conservation in the 21st Century*. Washington: Island Press, 2004.
[6]   McKibben, Bill. *The End of Nature*. New York: Random House, 1989.
[7]   Brower, David and Steve Chapple. *Let the Mountains Talk, Let the Rivers Run: A Call to Those Who Would Save the Earth*. San Francisco, CA: HarperCollins West, 1995.
[8]   Dudley, William (Editor). *Biodiversity*. San Diego, CA: Greenhouse Press, 2002.
[9]   Glavin, Terry. *The Sixth Extinction: Journeys Among the Lost and Left Behind*. Thomas Dunne Books/St. Martin's Press, 2007.
[10] Dudley, *Biodiversity*.
[11] Glavin, *The Sixth Extinction*.
[12] Ibid.
[13] Terborgh, John. *Requiem for Nature*. Washington, D.C.: Island Press, 1999.
[14] Glavin, *The Sixth Extinction*.
[15] Butler, Tom. *Wild Earth: Wild Ideas for a World Out Of Balance*. Minneapolis, MN: Milkweed Editions, 2002.
[16] Sussman, Art. *Dr. Art's Guide to Planet Earth: For Earthlings Ages 12 to 120*. White River Junction, VT: Chelsea Green Pub. Co., 2000.
[17] Chivian, Eric and Aaron Bernstein. *Sustaining Life: How Human Health Depends On Biodiversity*. Oxford, New York: Oxford University Press, 2008.

CORAL

[1] Kolbert, Elizabeth. *The Sixth Extinction: An Unnatural History*. New York: Henry Holt and Company, 2014.

[2] Sapp, Jan. *What Is Natural? Coral Reef Crisis*. New York: Oxford University Press, 1999.

[3] Dudley, William (Editor). *Biodiversity*. San Diego, CA: Greenhouse Press, 2002.

[4] Kolbert, *The Sixth Extinction*.

BACTERIA

[1] Wilson, E.O. *The Creation: An Appeal to Save Life on Earth*. New York: Norton, 2006.

[2] Margulis, Lynn and Karlene Schwartz. *Five Kingdoms: An Illustrated Guide to the Phyla of Life on Earth*. New York: W.H. Freeman, 1998.

COSMOS

[1] Ferris, Timothy. *First Glimpse at the Hidden Cosmos*. National Geographic, Vol. 227, No. 1, p. 108-123, January, 2015.

[2] *Cosmos: A Spacetime Odyssey*. Duryan, Ann, Seth McFarlane, Mitchell Cannold, Brannon Braga, Jason Clark, Neil deGrasse Tyson, and Alan Silvestri. Cosmos Studios, Inc.; Fuzzy Door Productions, 2013. Beverly Hills, CA: Twentieth Century Fox Home Entertainment, 2014. DVD.

[3] Devorkin, David H. and Robert W. Smith. *The Hubble Cosmos*. Washington, D.C.: National Geographic Society and Smithsonian Institute, 2015.

[4] Ibid.

[5] Weintraub, David. *How Old Is the Universe?* Princeton, NJ: Princeton University Press, 2011.

[6] Ibid.

[7] Devorkin, David H. *The Hubble Cosmos*.

[8] Weintraub, David. *How Old Is the Universe?*

9 Devorkin, David H. *The Hubble Cosmos*.

10 Ibid.

DOMINANT CULTURAL IDEOLOGY (DCI, DC, DIC, DI)

1 McKibben, Bill. *The End of Nature*. New York: Random House, 1989.

2 Devall, Bill. *Clearcut: The Tragedy of Industrial Forestry*. San Francisco: SierraClub Books/Earth Island Press, 1994.

3 Berry, Thomas. *The Great Work: Our Way into the Future*. New York: Bell Tower, 1999.

4 Hedges, Chris. *The World As It Is: Dispatches on the Myth of Human Progress*. New York: Nation Books, 2010.

5 Wilson, E.O. *The Creation: An Appeal to Save Life on Earth*. New York: Norton, 2006.

6 Hedges, *The World As It Is*.

7 Ehrenfeld, David. *The Arrogance of Humanism*. New York: Oxford University Press, 1978.

8 Mason, Jim. *An Unnatural Order: Why We Are Destroying the Planet and Each Other*. New York: Continuum, 1997.

9 Butler, Tom. *Wild Earth: Wild Ideas for a World Out Of Balance*. Minneapolis, MN: Milkweed Editions, 2002.

10 Mason, *An Unnatural Order*.

11 Wilson, *The Creation*.

12 Brower, David and Steve Chapple. *Let the Mountains Talk, Let the Rivers Run: A Call to Those Who Would Save the Earth*. San Francisco, CA: HarperCollins West, 1995.

13 Ehrenfeld, *The Arrogance of Humanism*.

14 Carns, Ted. *Off On Our Own: Living Off-Grid in Comfortable Independence*. Pittsburgh: St. Lynn's Press, 2011.

15 Hedges, *The World As It Is*.

16 Mason, *An Unnatural Order*.

17 Devall, *Clearcut*.

EVOLUTION

[1] Chivian, Eric and Aaron Bernstein. *Sustaining Life: How Human Health Depends On Biodiversity.* Oxford, New York: Oxford University Press, 2008.

[2] Suarez, Daniel. *Daemon.* New York: Dutton, 2009.

[3] Margulis, Lynn. *Symbiotic Planet: A New Look of Evolution.* New York: Basic Books, 1998.

[4] Ibid.

[5] Ibid.

[6] Margulis, Lynn and Karlene Schwartz. *Five Kingdoms: An Illustrated Guide to the Phyla of Life on Earth.* New York: W.H. Freeman, 1998.

[7] *Cosmos: A Spacetime Odyssey.* Duryan, Ann, Seth McFarlane, Mitchell Cannold, Brannon Braga, Jason Clark, Neil deGrasse Tyson, and Alan Silvestri. Cosmos Studios, Inc.; Fuzzy Door Productions, 2013. Beverly Hills, CA: Twentieth Century Fox Home Entertainment, 2014. DVD.

[8] Margulis, Lynn. *Five Kingdoms.*

[9] Pert, Candace. *Molecules of Emotion: Why You Feel the Way You Feel.* New York, New York: Simon & Schuster, Inc., 1997.

[10] grimwade.biochem.unimelb.edu.au/cone/main.html

OVERPOPULATION

[1] Ayres, Ed. *God's Last Offer: Negotiating for a Sustainable Future.* New York, NY: Four Walls Eight Windows, 1999.

[2] Korten, David. *When Corporations Rule the World.* West Hanford, Conn.: Kumarian Press; San Francisco, CA: Berrett Koehler Publications, 1995.

[3] Mason, Jim. *An Unnatural Order: Why We Are Destroying the Planet and Each Other.* New York: Continuum, 1997.

PREVIOUS EXTINCTIONS

1   Kump, Lee, James Kasting, and Robert Crane. *The Earth System*. Upper Saddle River, N.J.: Pearson Prentice Hall, 2004.
2   Foreman, Dave. *Rewilding North America: A Vision for Conservation in the 21ˢᵗ Century*. Washington: Island Press, 2004.
3   Corwin, Jeff. *100 Heartbeats: The Race to Save Earth's Most Endangered Species*. New York, New York: Rodale, 2009.
4   Wilson, E.O. *The Creation: An Appeal to Save Life on Earth*. New York: Norton, 2006.
5   McGavin, George. *Endangered: Wildlife on the Brink of Extinction*. Buffalo, N.Y.; Richmond Hill, Ont.: Firefly Books, 2006.

SPECIES TYPES

1   McGavin, George. *Endangered: Wildlife on the Brink of Extinction*. Buffalo, N.Y.; Richmond Hill, Ont.: Firefly Books, 2006.

HOTSPOTS

1   Glavin, Terry. *The Sixth Extinction: Journeys Among the Lost and Left Behind*. Thomas DunneBooks/St. Martin's Press, 2007.
2   Mittenmeier, Russell. *Hotspots Revisited*. Mexico City, Mexico: Cemex, 2004.

EARTH SYSTEMS

1   McKibben, Bill. *The End of Nature*. New York: Random House, 1989.
2   Holden, Joseph. *Physical Geography: the Basics*. New York: Routledge, 2011.
3   Flannery, Tim. *Now or Never: Why We Must Act Now to End Climate Change and Create a Sustainable Future*. New York: Atlantic Monthly Press; Distributed by Publisher's Group West, 2009.

FORESTS

[1] Margulis, Lynn. *Symbiotic Planet: A New Look of Evolution*. New York: Basic Books, 1998.
[2] Kump, Lee, James Kasting, and Robert Crane. *The Earth System*. Upper Saddle River, N.J.: Pearson Prentice Hall, 2004.
[3] Terborgh, John. *Requiem for Nature*. Washington, D.C.: Island Press, 1999.
[4] Ibid.
[5] Ibid.
[6] Liddick, Donald. *Crimes Against Nature: Illegal Industries and the Global Environment*. Santa Barbara, CA: Praeger, 2011.
[7] London, Mark and Brian Kelly. *The Last Forest: The Amazon in the Age of Globalization*. New York: Random House, 2007.
[8] Ibid.
[9] Ibid.
[10] Wilson, E.O. *The Creation: An Appeal to Save Life on Earth*. New York: Norton, 2006.
[11] Corwin, Jeff. *100 Heartbeats: The Race to Save Earth's Most Endangered Species*. New York, New York: Rodale, 2009.
[12] McKibben, Bill. *The End of Nature*. New York: Random House, 1989.
[13] Buchmann, Stephen and Gary Nabhan. *The Forgotten Pollinators*. Washington, D.C.: Island Press, 1996.
[14] Meyer, Stephen. *The End of the Wild*. Somerville, Mass.: Boston Review; Cambridge, Mass.: MIT Press, 2006.
[15] Terborgh, *Requiem for Nature*.
[16] Fraser, Caroline. *Rewilding the World: Dispatches from the Conservation Revolution*. New York: Metropolitan Books, 2009.
[17] McGavin, George. *Endangered: Wildlife on the Brink of Extinction*. Buffalo, N.Y.; Richmond Hill, Ont.: Firefly Books, 2006.
[18] Wilson, *The Creation*.
[19] Preston, Richard. *The Wild Trees: A Story of Passion and Daring*. New York: Random House, 2007.
[20] Foreman, Dave. *Rewilding North America: A Vision for Conservation in the 21st Century*. Washington: Island Press, 2004.

21 Devall, Bill. *Clearcut: The Tragedy of Industrial Forestry*. San Francisco: SierraClub Books/Earth Island Press, 1994.

22 Laurance, William. *Stinging Trees and Wait-a-Whiles: Confessions of a Rainforest Biologist*. Chicago, IL: University of Chicago Press, 2000.

23 Devall, *Clearcut*.

24 *Who Bombed Judi Bari?* Mary Thomson, Liz and Darryl Cheney. Hokey Pokey Productions, 2012. Audio Book or DVD.

25 Chadwick, Douglas. *The Truth About Tongass*. National Geographic Magazine. July 2007. ngm.nationalgeographic.com/2007/07/tongass/chadwick-text/7

26 Devall, *Clearcut*.

27 Ibid.

28 Ibid.

29 Butler, Tom. *Wild Earth: Wild Ideas for a World Out Of Balance*. Minneapolis, MN: Milkweed Editions, 2002.

30 Preston, *The Wild Trees*.

31 Devall, *Clearcut*.

32 Liddick, *Crimes Against Nature*.

33 Devall, *Clearcut*.

34 Ibid.

35 Ibid.

36 Ibid.

37 biology.duke.edu/bio265/jlp13/myco.php?t=nutrient

38 Preston, *The Wild Trees*.

39 Ibid.

ISLANDIZATION

1 Glavin, Terry. *The Sixth Extinction: Journeys Among the Lost and Left Behind*. Thomas Dunne Books/St. Martin's Press, 2007.

2 Kolbert, Elizabeth. *The Sixth Extinction: An Unnatural History*. New York: Henry Holt and Company, 2014.

3 Meyer, Stephen. *The End of the Wild*. Somerville, Mass.: Boston Review; Cambridge, Mass.: MIT Press, 2006.

4   Chivian, Eric and Aaron Bernstein. *Sustaining Life: How Human Health Depends On Biodiversity.* Oxford, New York: Oxford University Press, 2008.

5   Fraser, Caroline. *Rewilding the World: Dispatches from the Conservation Revolution.* New York: Metropolitan Books, 2009.

PRIMATES

1   Corwin, Jeff. *100 Heartbeats: The Race to Save Earth's Most Endangered Species.* New York, New York: Rodale, 2009.

2   Chivian, Eric and Aaron Bernstein. *Sustaining Life: How Human Health Depends On Biodiversity.* Oxford, New York: Oxford University Press, 2008.

NATURE DEFINED

1   Butler, Tom. *Wild Earth: Wild Ideas for a World Out Of Balance.* Minneapolis, MN: Milkweed Editions, 2002.

HISTORY OF CONSERVATION

1   Dudley, William (Editor). *Biodiversity.* San Diego, CA: Greenhouse Press, 2002.

2   Devall, Bill. *Clearcut: The Tragedy of Industrial Forestry.* San Francisco: SierraClub Books/Earth Island Press, 1994.

3   Foreman, Dave. *Rewilding North America: A Vision for Conservation in the 21ˢᵗ Century.* Washington: Island Press, 2004.

NORMALCY BIAS

1   Gage, Matilda Joslyn. *Woman, Church and State.* Amherst, New York: Humanity Books, 2002. Originally published by C.H. Kerr: Chicago, Illinois, 1893.

2   Greenberg, Paul. *Four Fish: The Future of the Last Wild Food.* Waterville, ME: Thorndike Press, 2010.

3   Quinn, Daniel. *Ishmael.* New York: Bantam/Turner Books, 1992.

DOMESTICATION

1   Goodall, Jane and Marc Beckoff. *The Ten Trusts: What We Must Do to Care for the Animals We Love.* San Francisco: Harper San Francisco, 2002.

2   Wuerthner, George. *Welfare Ranching: The Subsidized Destruction of the American West.* Washington, D.C.: Island Press, 2002.

3   *The Cove.* Psihoyos, Louie. Santa Monica, CA: Lion's Gate Entertainment, 2009. DVD.

4   *Racing Extinction.* Psihoyos, Louie. Seattle, WA: Vulcan Productions, 2015. DVD.

FRESH WATER

1   Sussman, Art. *Dr. Art's Guide to Planet Earth: For Earthlings Ages 12 to 120.* White River Junction, VT: Chelsea Green Pub. Co., 2000.

2   Dudley, William (Editor). *Biodiversity.* San Diego, CA: Greenhouse Press, 2002.

3   Chivian, Eric and Aaron Bernstein. *Sustaining Life: How Human Health Depends On Biodiversity.* Oxford, New York: Oxford University Press, 2008.

4   Goodall, Jane and Marc Beckoff. *The Ten Trusts: What We Must Do to Care for the Animals We Love.* San Francisco: Harper San Francisco, 2002.

5   Powell, James Lawrence. *Dead Pool: Lake Powell, Global Warming, and the Future of Water in the West.* Berkeley: University of California Press, 2008.

6   Wilson, E.O. *The Creation: An Appeal to Save Life on Earth.* New York: Norton, 2006.

7   Kump, Lee, James Kasting, and Robert Crane. *The Earth System.* Upper Saddle River, N.J.: Pearson Prentice Hall, 2004.

[8] Powell, *Dead Pool.*

[9] Ibid.

[10] Ibid.

[11] Ibid.

[12] Ibid.

[13] mead.uslakes.info/Level.asp

[14] lakepowell.water-data.com

[15] Pearce, Fred. *When the Rivers Run Dry: Water, the Defining Crisis of the Twenty-first Century.* Boston: Beacon Press, 2006.

ARTIFICIALITY

[1] Carns, Ted. *Off On Our Own: Living Off-Grid in Comfortable Independence.* Pittsburgh: St. Lynn's Press, 2011.

[2] De Becker, Gavin. *The Gift of Fear: Survival Signals that Protect Us from Violence.* New York, New York: Dell Pub., 1998.

[3] Meyer, Stephen. *The End of the Wild.* Somerville, Mass.: Boston Review; Cambridge, Mass.: MIT Press, 2006.

[4] Butler, Tom. *Wild Earth: Wild Ideas for a World Out Of Balance.* Minneapolis, MN: Milkweed Editions, 2002.

[5] Glavin, Terry. *The Sixth Extinction: Journeys Among the Lost and Left Behind.* Thomas Dunne Books/St. Martin's Press, 2007.

[6] Lambin, Eric. *The Middle Path: Avoiding Environmental Catastrophe.* Chicago University Press, 2007.

ECONOMICS

[1] Korten, David. *When Corporations Rule the World.* West Hanford, Conn.: Kumarian Press; San Francisco, CA: Berrett Koehler Publications, 1995.

[2] Ibid.

[3] Ibid.

[4] Faux, Geoffrey. *The Global Class War: How America's Bipartisan Elite Lost Our Future – and What It Will Take to Win It Back.* Hoboken, N.J.: Wiley, 2006.

[5] Deloria Jr., Vine. *We Talk, You Listen: New Tribes, New Turf.* New York: MacMillan, 1970.

[6] Liddick, Donald. *Crimes Against Nature: Illegal Industries and the Global Environment.* Santa Barbara, CA: Praeger, 2011.

[7] Ehrenfeld, David. *The Arrogance of Humanism.* New York: Oxford University Press, 1978.

[8] Brower, David and Steve Chapple. *Let the Mountains Talk, Let the Rivers Run: A Call to Those Who Would Save the Earth.* San Francisco, CA: HarperCollins West, 1995.

[9] Lambin, Eric. *The Middle Path: Avoiding Environmental Catastrophe.* Chicago University Press, 2007.

[10] Korten, *When Corporations Rule the World.*

[11] Ibid.

RHINOCEROSES

[1] Corwin, Jeff. *100 Heartbeats: The Race to Save Earth's Most Endangered Species.* New York, New York: Rodale, 2009.

[2] Kolbert, Elizabeth. *The Sixth Extinction: An Unnatural History.* New York: Henry Holt and Company, 2014.

EVOLUTION OF *HOMO SAPIENS*

[1] Kolbert, Elizabeth. *The Sixth Extinction: An Unnatural History.* New York: Henry Holt and Company, 2014.

[2] McDougall, Christopher. *Born To Run: A Hidden Tribe, Superathletes, and the Greatest Race the World Has Never Seen.* New York: Alfred A. Knopf, 2009.

INSECTS

[1] Wilson, E.O. *The Creation: An Appeal to Save Life on Earth.* New York: Norton, 2006.

[2] Ibid.

[3] Ibid.

[4]   Ibid.

CORPORATIONS

[1]   Williams, Terry Tempest. *The Open Space of Democracy.* Great Barrington, MA: The Orion Society, 2004.
[2]   Korten, David. *When Corporations Rule the World.* West Hanford, Conn.: Kumarian Press; San Francisco, CA: Berrett Koehler Publications, 1995.
[3]   Ibid.
[4]   Berry, Thomas. *The Great Work: Our Way into the Future.* New York: Bell Tower, 1999.
[5]   Korten, *When Corporations Rule the World.*
[6]   Berry, *The Great Work.*
[7]   Faux, Geoffrey. *The Global Class War: How America's Bipartisan Elite Lost Our Future – and What It Will Take to Win It Back.* Hoboken, N.J.: Wiley, 2006.
[8]   Korten, *When Corporations Rule the World.*
[9]   Berry, *The Great Work.*
[10] Korten, *When Corporations Rule the World.*
[11] Berry, *The Great Work.*
[12] Ibid.
[13] Suarez, Daniel. *Freedom™.* New York: Dutton, 2010.
[14] MacDonald, Christine. *Green, Inc.: An Environmental Insider Reveals How a Good Cause Has Gone Bad.* Guilford, CT: Lyons Press, an imprint of The Globe Pequot Press, 2008.
[15] *Cowspiracy.* Kip Anderson and Keegan Kuhn. A.U.M. Films & First Spark Media, 2014. DVD.
[16] Kolbert, Elizabeth. *The Sixth Extinction: An Unnatural History.* New York: Henry Holt and Company, 2014.
[17] Carns, Ted. *Off On Our Own: Living Off-Grid in Comfortable Independence.* Pittsburgh: St. Lynn's Press, 2011.
[18] Korten, *When Corporations Rule the World.*
[19] Ibid.
[20] Ibid.

[21] Hedges, Chris. *The World As It Is: Dispatches on the Myth of Human Progress*. New York: Nation Books, 2010.

[22] Korten, *When Corporations Rule the World*.

[23] Ibid.

[24] Ayres, Ed. *God's Last Offer: Negotiating for a Sustainable Future*. New York, NY: Four Walls Eight Windows, 1999.

[25] Berry, *The Great Work*.

[26] Korten, *When Corporations Rule the World*.

[27] Ibid.

[28] Faux, *The Global Class War*.

[29] Korten, *When Corporations Rule the World*.

[30] Ibid.

[31] Hedges, *The World As It Is*.

[32] Ibid.

[33] Korten, *When Corporations Rule the World*.

DENIAL

[1] *Chasing Ice*. Orlowski, Jeff. New York: Docurama, Cinedigm Entertainment, 2013. DVD.

[2] Suarez, Daniel. *Freedom™*. New York: Dutton, 2010.

[3] *League of Denial; the NFL's Concussion Crisis*. Kirk, Michael and Mike Wiser, Steve Fainaru, and Mark Fainaru-Wada. Kirk Doc. Group PBS, 2013. DVD.

[4] *Deliver Us from Evil*. Berg, Amy. Santa Monica, CA: Lion's Gate Entertainment, 2006. DVD.

CLIMATE CHANGE

[1] Pearson, Richard and American Museum of Natural History. *Driven To Extinction: The Impact of Climate Change on Biodiversity*. New York: Sterling, 2011.

[2] Balouet, Jean-Christophe, Eric Albert, and Joan Robb. *Extinct Species of the World*. New York: Barron's, 1990.

3   Kolbert, Elizabeth. *The Sixth Extinction: An Unnatural History.* New York: Henry Holt and Company, 2014.

4   Powell, James Lawrence. *Dead Pool: Lake Powell, Global Warming, and the Future of Water in the West.* Berkeley: University of California Press, 2008.

5   Fraser, Caroline. *Rewilding the World: Dispatches from the Conservation Revolution.* New York: Metropolitan Books, 2009.

6   ncdc.noaa.gov

7   Kolbert, *The Sixth Extinction.*

8   Powell, James Lawrence. *Dead Pool.*

9   Chivian, Eric and Aaron Bernstein. *Sustaining Life: How Human Health Depends On Biodiversity.* Oxford, New York: Oxford University Press, 2008.

10   Ibid.

11   Smith, Laurence. *The World in 2050: Four Forces Shaping Civilization's Northern Future.* New York, N.Y.: Plume, 2011.

12   Chivian, *Sustaining Life.*

13   ncdc.noaa.gov

14   Flannery, Tim. *Now or Never: Why We Must Act Now to End Climate Change and Create a Sustainable Future.* New York: Atlantic Monthly Press; Distributed by Publisher's Group West, 2009.

15   Kump, Lee, James Kasting, and Robert Crane. *The Earth System.* Upper Saddle River, N.J.: Pearson Prentice Hall, 2004.

16   *Gasland Part II.* Josh Fox. New York: Docurama, Cinedigm Entertainment, 2014. DVD.

17   Pearson, *Driven To Extinction.*

ENERGY

1   Lovelock, James. *The Vanishing Face of Gaia: A Final Warning.* New York: Basic Books, 2009.

2   *Gas Hole.* Wagener, Jeremy and Scott D. Roberts. Burbank, CA: Cinema Libre Studios, 2008. DVD.

3   *Thrive: What On Earth Will It Take?* Soquel, CA: Clear Compas Media, 2011. DVD.

4   *Who Killed the Electric Car?* Culver City, CA: Sony Pictures, 2006. DVD.

5   Pimentel, David. "Corn Ethanol as Energy: The Case Against US Production." *Harvard International Review.* Vol. 31; pg 50-3. Summer 2009. Opposing Viewpoints Resource Center. Web. 12 August 2010.

6   Tyner, Wallace. "The US Biofuels Boom: Its Origins, Current Status, and Future Prospects." *Bioscience.* Vol. 58, Iss. 7; pg. 646-74. July/August 2008. Proquest. Web. 12 August 2010.

7   Johnson, Robbin, and Ford Runge. "Ethanol: Train Wreck Ahead?" *Issues in Science and Technology.* Vol. 24; pg. 25-30. Fall 2007. Opposing Viewpoints Resource Center. Web. 12 August 2010.

8   Oliveira, Ana Christina, and Luisa Gouveia. "Microalgae as a Raw Material for Biofuels Production." *Journal for Industrial Microbiological Biotechnology.* Vol. 36; Pg. 269-74. 4 November 2008. EBSCO. Web. 10 August 2010.

9   Pimentel, "Corn Ethanol as Energy."

10   Potera, Carol. "Corn Ethanol Goal Revives Dead Zone Concerns." *Environmental Health Perspectives.* Vol. 116; pg. 242-3. June 2008. Web. 12 August 2010.

11   Tyner, "The US Biofuels Boom."

12   Ibid.

13   Pimentel, "Corn Ethanol as Energy."

14   Ibid.

15   Ibid.

16   Ibid.

17   Wolinsky, Howard. "The Economics of Biofuels." *European Molecular Biology Organization Reports.* Vol. 10. No. 6; pg. 551-53. 2009. EBSCO. Web. 10 August 2010.

18   *Fuel.* Dir. Josh Tickell. Open Pictures. Blue Water Entertainment. 2008. DVD.

19   Bhat, Kiran. "Misplaced Priorities: Ethanol Promotion and its Unintended Consequences." *Harvard International Review.* Vol. 30; pg. 30-4. Spring 2008. Opposing Viewpoints Resource Center. Web. 12 August 2010.

20 Gouveia, Luisa. "Neochloris Oleabundans UTEX #1185: A Suitable Renewable Lipid Source for Biofuel Production." *Journal of Industrial Microbiological Biotechnology*. Vol. 36; pg. 821-26. 18 April 2009. EBSCO. Web. 10 August 2010.

21 Ibid.

22 Ibid.

23 Tyner, "The US Biofuels Boom."

24 Shuster, Joseph. *Beyond Fossil Fuels: the Roadmap to Energy Independence by 2040*. Beaver Pond Press, Edina, Minnesota. 2008.

25 Lane, Jim. "Jet Stream: Biofuels Digest Special Report on Aviation Biofuels." *biofuelsdigest*. Word Press. 3 March 2010. Web. 11 August 2010.

26 Oliveira, "Microalgae as a Raw Material for Biofuels Production."

27 Smith, Laurence. *The World in 2050: Four Forces Shaping Civilization's Northern Future*. New York, N.Y.: Plume, 2011.

28 Ibid.

RADIOACTIVITY

1 Caldicott, Helen. *Nuclear Power Is Not the Answer*. New York: New Press; London: Turnaround, 2006, 2007.

2 Ibid.

3 Ibid.

4 Ibid.

5 Ibid.

6 Ibid.

7 Lovelock, James. *The Vanishing Face of Gaia: A Final Warning*. New York: Basic Books, 2009.

8 Caldicott, *Nuclear Power Is Not the Answer*.

9 Ibid.

10 Ibid.

11 Ibid.

12 Ibid.

13 Ibid.

14 Ibid.

15 Ibid.

[16] http://enenews.com/ap-biggest-question-fukushimas-melted-fuel-asahi-fuel-mainichi-one-fuel-experts-nuclear-cores-hit-groundwater-could-melted-ground-world-never-case-like-fukushima-fuel-melted-fell-video. June, 23, 2015.

[17] Caldicott, *Nuclear Power Is Not the Answer.*

[18] Liddick, Donald. *Crimes Against Nature: Illegal Industries and the Global Environment.* Santa Barbara, CA: Praeger, 2011.

TIGERS

[1] Corwin, Jeff. *100 Heartbeats: The Race to Save Earth's Most Endangered Species.* New York, New York: Rodale, 2009.

[2] Glavin, Terry. *The Sixth Extinction: Journeys Among the Lost and Left Behind.* Thomas Dunne Books/St. Martin's Press, 2007.

[3] Liddick, Donald. *Crimes Against Nature: Illegal Industries and the Global Environment.* Santa Barbara, CA: Praeger, 2011.

[4] Foreman, Dave. *Rewilding North America: A Vision for Conservation in the 21st Century.* Washington: Island Press, 2004.

[5] Corwin, *100 Heartbeats.*

[6] Glavin, *The Sixth Extinction*

[7] Vaillant, John. *The Tiger: A True Story of Vengeance and Survival.* New York: Alfred A. Knopf, 2010.

[8] Corwin, *100 Heartbeats.*

HABITAT DESTRUCTION

[1] Dudley, William (Editor). *Biodiversity.* San Diego, CA: Greenhouse Press, 2002.

[2] Chivian, Eric and Aaron Bernstein. *Sustaining Life: How Human Health Depends On Biodiversity.* Oxford, New York: Oxford University Press, 2008.

[3] Goodall, Jane and Marc Beckoff. *The Ten Trusts: What We Must Do to Care for the Animals We Love.* San Francisco: Harper San Francisco, 2002.

[4]   Foreman, Dave. *Rewilding North America: A Vision for Conservation in the 21ˢᵗ Century.* Washington: Island Press, 2004.

[5]   Wuerthner, George. *Welfare Ranching: The Subsidized Destruction of the American West.* Washington, D.C.: Island Press, 2002.

[6]   Powell, James Lawrence. *Dead Pool: Lake Powell, Global Warming, and the Future of +Water in the West.* Berkeley: University of California Press, 2008.

[7]   Chivian, *Sustaining Life.*

[8]   Lambin, Eric. *The Middle Path: Avoiding Environmental Catastrophe.* Chicago University Press, 2007.

[9]   Synnott, Mark. *Sins of the Aral Sea.* National Geographic, Vol. 227, No. 6, p. 114-131, January, 2015.

[10]   Devall, Bill. *Clearcut: The Tragedy of Industrial Forestry.* San Francisco: SierraClub Books/Earth Island Press, 1994.

[11]   Ibid.

[12]   Foreman, *Rewilding North America.*

[13]   Ibid.

[14]   Lambin, *The Middle Path.*

[15]   Korten, David. *When Corporations Rule the World.* West Hanford, Conn.: Kumarian Press; San Francisco, CA: Berrett Koehler Publications, 1995.

INVASIVE SPECIES

[1]   Meyer, Stephen. *The End of the Wild.* Somerville, Mass.: Boston Review; Cambridge, Mass.: MIT Press, 2006.

[2]   Ibid.

[3]   Wuerthner, George. *Welfare Ranching: The Subsidized Destruction of the American West.* Washington, D.C.: Island Press, 2002.

[4]   Foreman, Dave. *Rewilding North America: A Vision for Conservation in the 21ˢᵗ Century.* Washington: Island Press, 2004.

[5]   Terborgh, John. *Requiem for Nature.* Washington, D.C.: Island Press, 1999.

[6]   Sapp, Jan. *What Is Natural? Coral Reef Crisis.* New York: Oxford University Press, 1999.

[7]   Foreman, *Rewilding North America.*

8   Terborgh, *Requiem for Nature.*

9   Foreman, *Rewilding North America.*

10   Kolbert, Elizabeth. *The Sixth Extinction: An Unnatural History.* New York: Henry Holt and Company, 2014.

11   Dudley, William (Editor). *Biodiversity.* San Diego, CA: Greenhouse Press, 2002.

ICE

1   Kolbert, Elizabeth. *The Sixth Extinction: An Unnatural History.* New York: Henry Holt and Company, 2014.

2   Pearce, Fred. *When the Rivers Run Dry: Water, the Defining Crisis of the Twenty-first Century.* Boston: Beacon Press, 2006.

3   Parenti, Christian. *Tropic of Chaos: Climate Change and the New Geography of Violence.* New York: Nation Books, 2011.

4   Lovelock, James. *The Vanishing Face of Gaia: A Final Warning.* New York: Basic Books, 2009.

5   *Chasing Ice.* Orlowski, Jeff. New York: Docurama, Cinedigm Entertainment, 2013. DVD.

DESERTIFICATION

1   Ehrenfeld, David. *The Arrogance of Humanism.* New York: Oxford University Press, 1978.

2   Kolbert, Elizabeth. *The Sixth Extinction: An Unnatural History.* New York: Henry Holt and Company, 2014.

3   Mason, Jim. *An Unnatural Order: Why We Are Destroying the Planet and Each Other.* New York: Continuum, 1997.

4   Terborgh, John. *Requiem for Nature.* Washington, D.C.: Island Press, 1999.

5   Ibid.

6   Wuerthner, George. *Welfare Ranching: The Subsidized Destruction of the American West.* Washington, D.C.: Island Press, 2002.

7   Ibid.

8   Terborgh, *Requiem for Nature.*

⁹ Pearce, Fred. *When the Rivers Run Dry: Water, the Defining Crisis of the Twenty-first Century.* Boston: Beacon Press, 2006.

OCEAN

¹ Chivian, Eric and Aaron Bernstein. *Sustaining Life: How Human Health Depends On Biodiversity.* Oxford, New York: Oxford University Press, 2008.

² Kolbert, Elizabeth. *The Sixth Extinction: An Unnatural History.* New York: Henry Holt and Company, 2014.

³ Kump, Lee, James Kasting, and Robert Crane. *The Earth System.* Upper Saddle River, N.J.: Pearson Prentice Hall, 2004.

⁴ Chivian, *Sustaining Life.*

FISHERIES

¹ Glavin, Terry. *The Sixth Extinction: Journeys Among the Lost and Left Behind.* Thomas Dunne Books/St. Martin's Press, 2007.

² Greenberg, Paul. *Four Fish: The Future of the Last Wild Food.* Waterville, ME: Thorndike Press, 2010.

³ *End of the Line.* Rupert Murray and Ted Dansen, Claire Lewis, George Duffield, and Charles Clover. New York: Docurama, 2009. DVD.

⁴ Greenberg, *Four Fish.*

⁵ *End of the Line.*

⁶ Greenberg, *Four Fish.*

⁷ Ibid.

⁸ Ibid.

⁹ Ibid.

¹⁰ Chivian, Eric and Aaron Bernstein. *Sustaining Life: How Human Health Depends On Biodiversity.* Oxford, New York: Oxford University Press, 2008.

¹¹ Liddick, Donald. *Crimes Against Nature: Illegal Industries and the Global Environment.* Santa Barbara, CA: Praeger, 2011.

¹² Glavin, *The Sixth Extinction.*

13 Ibid.

14 Chivian, *Sustaining Life*.

15 Goodall, Jane and Marc Beckoff. *The Ten Trusts: What We Must Do to Care for the Animals We Love*. San Francisco: Harper San Francisco, 2002.

16 Glavin, *The Sixth Extinction*.

ANIMALS

1 Wilson, E.O. *The Creation: An Appeal to Save Life on Earth*. New York: Norton, 2006.

2 Margulis, Lynn and Karlene Schwartz. *Five Kingdoms: An Illustrated Guide to the Phyla of Life on Earth*. New York: W.H. Freeman, 1998.

3 Ibid.

4 Liddick, Donald. *Crimes Against Nature: Illegal Industries and the Global Environment*. Santa Barbara, CA: Praeger, 2011.

5 Ibid.

6 Ibid.

7 Kolbert, Elizabeth. *The Sixth Extinction: An Unnatural History*. New York: Henry Holt and Company, 2014.

8 Liddick, *Crimes Against Nature: Illegal Industries and the Global Environment*.

9 Foreman, Dave. *Rewilding North America: A Vision for Conservation in the 21ˢᵗ Century*. Washington: Island Press, 2004.

10 Goodall, Jane and Marc Beckoff. *The Ten Trusts: What We Must Do to Care for the Animals We Love*. San Francisco: Harper San Francisco, 2002.

11 Ibid.

12 Foreman, *Rewilding North America*.

13 Corwin, Jeff. *100 Heartbeats: The Race to Save Earth's Most Endangered Species*. New York, New York: Rodale, 2009.

14 Glavin, Terry. *The Sixth Extinction: Journeys Among the Lost and Left Behind*. Thomas Dunne Books/St. Martin's Press, 2007.

RELIGION, PART I

1   Campbell, Joseph and Bill Moyers. *The Power of Myth*. New York: Doubleday, 1988.
2   Quinn, Daniel. *My Ishmael*. New York: Bantam Books, 1997.
3   Mason, Jim. *An Unnatural Order: Why We Are Destroying the Planet and Each Other*. New York: Continuum, 1997.
4   Quinn, Daniel. *Ishmael*. New York: Bantam/Turner Books, 1992.
5   Jensen, Derrick. *Endgame: The Problem of Civilization, Vol. 1*. New York: Seven Stories, 2006.
6   Manes, Christopher. *Green Rage: Radical Environmentalism and the Unmaking of Civilization*. Boston: Little, Brown, 1990.

WOLVES

1   Fraser, Caroline. *Rewilding the World: Dispatches from the Conservation Revolution*. New York: Metropolitan Books, 2009.
2   Ibid.

CATTLE

1   Jacobs, Lynn. *Waste of the West: Public Lands Ranching*. Tucson, AZ: L. Jacobs, 1991.
2   Wuerthner, George. *Welfare Ranching: The Subsidized Destruction of the American West*. Washington, D.C.: Island Press, 2002.
3   Chivian, Eric and Aaron Bernstein. *Sustaining Life: How Human Health Depends On Biodiversity*. Oxford, New York: Oxford University Press, 2008.
4   Sussman, Art. *Dr. Art's Guide to Planet Earth: For Earthlings Ages 12 to 120*. White River Junction, VT: Chelsea Green Pub. Co., 2000.
5   Jacobs, Lynn. *Waste of the West*.
6   Foreman, Dave. *Rewilding North America: A Vision for Conservation in the 21ˢᵗ Century*. Washington: Island Press, 2004.
7   Ibid.
8   Wuerthner, *Welfare Ranching*.

9   Jacobs, Lynn. *Waste of the West.*

10 Wuerthner, *Welfare Ranching.*

11 Barker, Rodney. *And the Waters Turned to Blood: The Ultimate Biological Threat.* New York, New York: Simon & Schuster, 1997.

12 Wuerthner, *Welfare Ranching.*

13 Ibid.

14 Ibid.

15 Ibid.

16 Powell, James Lawrence. *Dead Pool: Lake Powell, Global Warming, and the Future of Water in the West.* Berkeley: University of California Press, 2008.

17 Wuerthner, *Welfare Ranching.*

18 Flannery, Tim. *Now or Never: Why We Must Act Now to End Climate Change and Create a Sustainable Future.* New York: Atlantic Monthly Press; Distributed by Publisher's Group West, 2009.

19 Wuerthner, *Welfare Ranching.*

20 Ibid.

21 Foreman, *Rewilding North America.*

22 Wuerthner, *Welfare Ranching.*

23 Ibid.

24 Ibid.

25 Jacobs, Lynn. *Waste of the West.*

26 Wuerthner, *Welfare Ranching.*

27 Jacobs, Lynn. *Waste of the West.*

28 Wuerthner, *Welfare Ranching.*

29 Ibid.

30 Ibid.

31 Ibid.

32 Ibid.

33 Jacobs, Lynn. *Waste of the West.*

34 Wuerthner, *Welfare Ranching.*

35 Goodall, Jane and Marc Beckoff. *The Ten Trusts: What We Must Do to Care for the Animals We Love.* San Francisco: Harper San Francisco, 2002.

36 Chivian, *Sustaining Life.*

37 Goodall, *The Ten Trusts.*

38 Wuerthner, *Welfare Ranching.*
39 Lyman, Howard and Glen Merzer. *Mad Cowboy.: Plain Truth from the Cattle Rancher Who Won't Eat Meat.* New York: Schribner, 1998.
40 Jacobs, Lynn. *Waste of the West.*
41 Wuerthner, *Welfare Ranching.*
42 London, Mark and Brian Kelly. *The Last Forest: The Amazon in the Age of Globalization.* New York: Random House, 2007.
43 Meyer, Stephen. *The End of the Wild.* Somerville, Mass.: Boston Review; Cambridge, Mass.: MIT Press, 2006.
44 Terborgh, John. *Requiem for Nature.* Washington, D.C.: Island Press, 1999.
45 London, Mark and Brian Kelly. *The Last Forest: The Amazon in the Age of Globalization.*
46 Wuerthner, *Welfare Ranching.*
47 Chivian, *Sustaining Life.*
48 Wuerthner, *Welfare Ranching.*
49 Jacobs, Lynn. *Waste of the West.*
50 Ibid.

PRAIRIE DOGS

1 Wuerthner, George. *Welfare Ranching: The Subsidized Destruction of the American West.* Washington, D.C.: Island Press, 2002.
2 biologicaldiversity.org/news/press_releases/2015/wildlife-services-04-13-2015.html
3 Fraser, Caroline. *Rewilding the World: Dispatches from the Conservation Revolution.* New York: Metropolitan Books, 2009.
4 Ibid.

RELIGION, PART II

1 Gage, Matilda Joslyn. *Woman, Church and State.* Amherst, New York: Humanity Books, 2002. Originally published by C.H. Kerr: Chicago, Illinois, 1893.
2 Ibid.

3   Ibid.

4   Encyclopaedia Britannica, 15th Edition. *The Inquisition.* Vol. 6: 328:2b. Chicago: Encyclopaedia Britannica, 2007.

5   Gage, *Woman, Church and State.*

6   Shepard, Paul. *Man in the Landscape: A Historic View of the Esthetics of Nature.* College Station: Texas A&M University Press, 1991.

7   Butler, Tom. *Wild Earth: Wild Ideas for a World Out Of Balance.* Minneapolis, MN: Milkweed Editions, 2002.

8   Ibid.

9   Gage, *Woman, Church and State.*

10   Chivian, Eric and Aaron Bernstein. *Sustaining Life: How Human Health Depends On Biodiversity.* Oxford, New York: Oxford University Press, 2008.

11   Gage, *Woman, Church and State.*

12   Ibid.

13   Ibid.

14   Ibid.

15   Ibid.

16   Ibid.

17   Ibid.

18   Ibid.

19   Ibid.

20   Ibid.

21   Ibid.

22   Ehrenfeld, David. *The Arrogance of Humanism.* New York: Oxford University Press, 1978.

23   Gage, *Woman, Church and State.*

24   Deloria Jr., Vine. *We Talk, You Listen: New Tribes, New Turf.* New York: MacMillan, 1970.

25   Gage, *Woman, Church and State.*

26   *The End of Poverty?* Portello, Beth, Phillippe Diaz, and Martin Sheen. [Canoga Park], CA: Cinema Libre Studio, 2010. DVD.

27   Gage, *Woman, Church and State.*

28   Ibid.

OBSOLETE

1   Jensen, Derrick. *Endgame: The Problem of Civilization, Vol. 1.* New York: Seven Stories, 2006.
2   Wuerthner, George. *Welfare Ranching: The Subsidized Destruction of the American West.* Washington, D.C.: Island Press, 2002.
3   *Cosmos: A Spacetime Odyssey.* Duryan, Ann, Seth McFarlane, Mitchell Cannold, Brannon Braga, Jason Clark, Neil deGrasse Tyson, and Alan Silvestri. Cosmos Studios, Inc.; Fuzzy Door Productions, 2013. Beverly Hills, CA: Twentieth Century Fox Home Entertainment, 2014. DVD.
4   *Evolutionary Christianity.* Michael Dowd. Evolutionarychristianity. org, 2011-2015. DVD.
5   Dowd, Michael. *Thank God for Evolution: How the Marriage of Science and Religion Will Transform Your Life and Our World.* Tulsa OK: Council Oak Books, 2007.

RELIGION, PART III

1   Williams, Paul L. *The Vatican Exposed: Money, Murder, and the Mafia.* Amherst, New York: Prometheus Books, 2003.
2   Posner, Gerald. *God's Bankers: A History of Money and Power at the Vatican.* NY, NY: Simon & Schuster, 2015.
3   Williams, Paul L. *The Vatican Exposed.*
4   Posner, Gerald. *God's Bankers.*
5   Williams, Paul L. *The Vatican Exposed.*
6   Posner, Gerald. *God's Bankers.*
7   Ibid. (quoted from Peter C. Kent, *The Pope and the Duce: The International Impact of the Lateran Agreements* New York: St. Martins, 1981.)
8   Williams, Paul L. *The Vatican Exposed.*
9   Ibid.
10   Posner, Gerald. *God's Bankers.*
11   Ibid.
12   Williams, Paul L. *The Vatican Exposed.*
13   Ibid.
14   Posner, Gerald. *God's Bankers.*

15. Williams, Paul L. *The Vatican Exposed.*
16. Posner, Gerald. *God's Bankers.*
17. Williams, Paul L. *The Vatican Exposed.*
18. D'Antonio, Michael. *Mortal Sins: Sex, Crime, and the Era of Catholic Scandal.* New York: Thorndike Press, a part of Gale, Cengage Learning, 2013.
19. Posner, Gerald. *God's Bankers.*
20. Williams, Paul L. *The Vatican Exposed.*
21. Posner, Gerald. *God's Bankers.*
22. Ibid.
23. Hammer, Richard. *The Vatican Connection: The Astonishing Account of a Billion-Dollar Counterfeit Stock Deal Between the Mafia and the Catholic Church.* New York: Holt, Rinehart, and Winston, 1982.
24. Posner, Gerald. *God's Bankers.*
25. Hammer, Richard. *The Vatican Connection.*
26. Ibid.
27. Williams, Paul L. *The Vatican Exposed.*
28. Ibid.
29. Hammer, Richard. *The Vatican Connection.*
30. Williams, Paul L. *The Vatican Exposed.*
31. Posner, Gerald. *God's Bankers.*
32. D'Antonio, Michael. *Mortal Sins.*
33. Ibid.
34. *Deliver Us from Evil.* Berg, Amy. Santa Monica, CA: Lion's Gate Entertainment, 2006. DVD.
35. D'Antonio, Michael. *Mortal Sins.*
36. Ibid.
37. Posner, Gerald. *God's Bankers.*
38. D'Antonio, Michael. *Mortal Sins.*
39. Posner, Gerald. *God's Bankers.*
40. Ibid.
41. Ibid.

BIRDS

1   McGavin, George. *Endangered: Wildlife on the Brink of Extinction.* Buffalo, N.Y.; Richmond Hill, Ont.: Firefly Books, 2006.
2   Ibid.
3   Wuerthner, George. *Welfare Ranching: The Subsidized Destruction of the American West.* Washington, D.C.: Island Press, 2002.
4   Franzen, Jonathan. *Last Song for Migrating Birds.* National Geographic, July, 2013. http://ngm.nationalgeographic.com/2013/07/songbird-migration/franzen-text
5   *Emptying the Skies.* Kass, Douglas and Roger. Chicago: Music Box Films, 2013. DVD.
6   McGavin, George. *Endangered.*
7   Liddick, Donald. *Crimes Against Nature: Illegal Industries and the Global Environment.* Santa Barbara, CA: Praeger, 2011.
8   McGavin, George. *Endangered.*

FARMING

1   Mason, Jim. *An Unnatural Order: Why We Are Destroying the Planet and Each Other.* New York: Continuum, 1997.
2   Ibid.
3   Ibid.
4   Berry, Thomas. *The Great Work: Our Way into the Future.* New York: Bell Tower, 1999.
5   Lambin, Eric. *The Middle Path: Avoiding Environmental Catastrophe.* Chicago University Press, 2007.
6   Glavin, Terry. *The Sixth Extinction: Journeys Among the Lost and Left Behind.* Thomas Dunne Books/St. Martin's Press, 2007.
7   Ibid.
8   Pearce, Fred. *When the Rivers Run Dry: Water, the Defining Crisis of the Twenty-first Century.* Boston: Beacon Press, 2006.
9   Lambin, *The Middle Path.*
10 Pearce, *When the Rivers Run Dry.*
11 Smith, Laurence. *The World in 2050: Four Forces Shaping Civilization's Northern Future.* New York, N.Y.: Plume, 2011.

[12] Chivian, Eric and Aaron Bernstein. *Sustaining Life: How Human Health Depends On Biodiversity*. Oxford, New York: Oxford University Press, 2008.

[13] Dudley, William (Editor). *Biodiversity*. San Diego, CA: Greenhouse Press, 2002.

[14] Chivian, *Sustaining Life*.

[15] Corwin, Jeff. *100 Heartbeats: The Race to Save Earth's Most Endangered Species*. New York, New York: Rodale, 2009.

[16] Suarez, Daniel. *Freedom™*. New York: Dutton, 2010.

NATURAL SERVICES

[1] Chivian, Eric and Aaron Bernstein. *Sustaining Life: How Human Health Depends On Biodiversity*. Oxford, New York: Oxford University Press, 2008.

[2] Ibid.

[3] Wilson, E.O. *The Creation: An Appeal to Save Life on Earth*. New York: Norton, 2006.

POLLINATORS

[1] Preston, Richard. *The Wild Trees: A Story of Passion and Daring*. New York: Random House, 2007.

[2] Buchmann, Stephen and Gary Nabhan. *The Forgotten Pollinators*. Washington, D.C.: Island Press, 1996.

[3] Ibid.

[4] Ibid.

[5] Ibid.

[6] Butler, Tom. *Wild Earth: Wild Ideas for a World Out Of Balance*. Minneapolis, MN: Milkweed Editions, 2002.

[7] Buchmann, *The Forgotten Pollinators*.

[8] Ibid.

[9] Ibid.

[10] Ibid.

[11] Ibid.

[12] Ibid.

DEBT

[1] Lambin, Eric. *The Middle Path: Avoiding Environmental Catastrophe.* Chicago University Press, 2007.

NATIVE AMERICANS

[1] Carns, Ted. *Off On Our Own: Living Off-Grid in Comfortable Independence.* Pittsburgh: St. Lynn's Press, 2011.

[2] Hughes, J. Donald. *American Indian Ecology.* El Paso: Texas Western Press, 1983.

[3] Carns, *Off On Our Own.*

[4] Terborgh, John. *Requiem for Nature.* Washington, D.C.: Island Press, 1999.

[5] London, Mark and Brian Kelly. *The Last Forest: The Amazon in the Age of Globalization.* New York: Random House, 2007.

[6] Chivian, Eric and Aaron Bernstein. *Sustaining Life: How Human Health Depends On Biodiversity.* Oxford, New York: Oxford University Press, 2008.

[7] London, *The Last Forest.*

[8] Deloria Jr., Vine. *We Talk, You Listen: New Tribes, New Turf.* New York: MacMillan, 1970.

[9] Hughes, *American Indian Ecology.*

[10] Diamond, Jared. *The Third Chimpanzee: The Evolution and Future of the Human Animal.* New York: HarperPerennial, 1992.

[11] Butler, Tom. *Wild Earth: Wild Ideas for a World Out Of Balance.* Minneapolis, MN: Milkweed Editions, 2002.

[12] Leopold, Aldo. *A Sand County Almanac.* New York: Oxford University Press, 1949.

[13] Goodall, Jane and Marc Beckoff. *The Ten Trusts: What We Must Do to Care for the Animals We Love.* San Francisco: Harper San Francisco, 2002.

[14] Devall, Bill. *Clearcut: The Tragedy of Industrial Forestry*. San Francisco: SierraClub Books/Earth Island Press, 1994.

[15] Hughes, *American Indian Ecology*.

[16] Deloria Jr., *We Talk, You Listen*.

[17] Hughes, *American Indian Ecology*.

[18] McKibben, Bill. *The End of Nature*. New York: Random House, 1989.

[19] Hughes, *American Indian Ecology*.

[20] Ibid.

[21] Fraser, Caroline. *Rewilding the World: Dispatches from the Conservation Revolution*. New York: Metropolitan Books, 2009.

[22] Hughes, *American Indian Ecology*.

[23] Berry, Thomas. *The Great Work: Our Way into the Future*. New York: Bell Tower, 1999.

[24] Deloria Jr., *We Talk, You Listen*.

[25] Josephy Jr., Alvin. *Patriot Chiefs: A Chronicle of American Indian Leadership*. New York: Viking Press, 1961.

[26] Suarez, Daniel. *Freedom™*. New York: Dutton, 2010.

[27] Foreman, Dave. *Rewilding North America: A Vision for Conservation in the 21st Century*. Washington: Island Press, 2004.

ELEPHANTS

[1] Maisels, Fiona, et. al. *"Devastating Decline of Forest Elephants in Central Africa."* Plos ONE 8, no. 3 (March 2013): 1-13. Academic Search Premier, EBSCOhost, Feb. 13, 2014.

[2] Christy, Bryan. *Ivory Worship*. National Geographic Magazine. October 2012. ngm.nationalgeographic.com/2012/10/ivory/christy-text

IMPAIRMENT, SUPPRESSION, DISPLACEMENT (ISD)

[1] Wuerthner, George. *Thrillcraft: The Environmental Consequences of Motorized Recreation*. White River Junction, VT: Chelsea Green Pub. Co., 2007.

ARTIFICIAL INTELLIGENCE

1   Kurzweil, Ray. *The Age of Spiritual Machines: When Computers Exceed Human Intelligence.* New York: Penguin, 2000.

WHALES

1   Glavin, Terry. *The Sixth Extinction: Journeys Among the Lost and Left Behind.* Thomas Dunne Books/St. Martin's Press, 2007.
2   Ibid.
3   Greenberg, Paul. *Four Fish: The Future of the Last Wild Food.* Waterville, ME: Thorndike Press, 2010.

SOLUTIONS

1   Goodall, Jane and Marc Beckoff. *The Ten Trusts: What We Must Do to Care for the Animals We Love.* San Francisco: Harper San Francisco, 2002.
2   Butler, Tom. *Wild Earth: Wild Ideas for a World Out Of Balance.* Minneapolis, MN: Milkweed Editions, 2002.
3   *The Internet's Own Boy: The Story of Aaron Swartz.* Knappenberger, Brian. Los Angeles, CA: Luminant Media, 2014. DVD.
4   Smith, Laurence. *The World in 2050: Four Forces Shaping Civilization's Northern Future.* New York, N.Y.: Plume, 2011.
5   Klein, Naomi. *This Changes Everything: Capitalism vs. the Climate.* New York: Simon & Schuster, 2014.
6   Devall, Bill. *Clearcut: The Tragedy of Industrial Forestry.* San Francisco: SierraClub Books/Earth Island Press, 1994.
7   Berry, Thomas. *The Great Work: Our Way into the Future.* New York: Bell Tower, 1999.
8   Gage, Matilda Joslyn. *Woman, Church and State.* Amherst, New York: Humanity Books, 2002. Originally published by C.H. Kerr: Chicago, Illinois, 1893.
9   Goodall, Jane. *The Ten Trusts.*

10 *Bidder 70*. Gage, Beth and George.Telluride, CO: Gage & Gage Productions; [S.I.]: Distributed by First Run Features, 2013. DVD.

11 Korten, David. *When Corporations Rule the World*. West Hanford, Conn.: Kumarian Press; San Francisco, CA: Berrett Koehler Publications, 1995.

12 Berry, Thomas. *The Great Work*.

13 Tait, Malcolm. *Going, Going, Gone? Animals and Plants On the Brink of Extinction and How You Can Help*. London, England: Think Publishing; New York: Sterling Publishing Co. Inc., 2006.

14 Goodall, Jane. *The Ten Trusts*.

15 Klein, Naomi. *This Changes Everything*.

16 Hedges, Chris. *The World As It Is: Dispatches on the Myth of Human Progress*. New York: Nation Books, 2010.

17 Korten, David. *When Corporations Rule the World*.

18 Dowie, Mark. *Losing Ground: American Environmentalism at the Close of the Twentieth Century*. Cambridge, Massachusetts: The MIT Press, 1996.

19 MacDonald, Christine. *Green, Inc.: An Environmental Insider Reveals How a Good Cause Has Gone Bad*. Guilford, CT: Lyons Press, an imprint of The Globe Pequot Press, 2008.

20 Terborgh, John. *Requiem for Nature*. Washington, D.C.: Island Press, 1999.

21 Gage, Matilda Joslyn. *Woman, Church and State*.

22 Associated Press. *"Dallas club gets permit to import dead rhino."* Arizona Republic, azcentral.com, March 29, 2015.

23 Klein, Naomi. *This Changes Everything*.

24 Brower, David and Steve Chapple. *Let the Mountains Talk, Let the Rivers Run: A Call to Those Who Would Save the Earth*. San Francisco, CA: HarperCollins West, 1995.

25 Ibid.

26 Dowd, Michael. *Thank God for Evolution: How the Marriage of Science and Religion Will Transform Your Life and Our World*. Tulsa OK: Council Oak Books, 2007.

27 *Evolutionary Christianity*. Michael Dowd. evolutionarychristianity.org, 2011-2015. DVD.

28 Corwin, Jeff. *100 Heartbeats: The Race to Save Earth's Most Endangered Species*. New York, New York: Rodale, 2009.

29 teslaenergy.com

30 Lyman, Howard and Glen Merzer. *Mad Cowboy: Plain Truth from the Cattle Rancher Who Won't Eat Meat*. New York: Schribner, 1998.

31 Korten, David. *When Corporations Rule the World*.

32 Smith, Laurence. *The World in 2050*.

33 Devall, Bill. *Clearcut*.

34 *Cosmos*. Sagan, Carl, Ann Duryan, Steve Soter, and Adrian Malone. [Volume 2], [Volume 2]. Studio City, CA: Cosmos Studios, 2000. DVD.

35 De Becker, Gavin. *The Gift of Fear: Survival Signals that Protect Us from Violence*. New York, New York: Dell Pub., 1998.

APPENDIX, SPECIES TYPES

1 www.iucnredlist.org

2 http://en.wikipedia.org/wiki/ Lists_of_IUCN_Red_List_Critically_Endangered_species. (Easier to identify.)

3 Meyer, Stephen. *The End of the Wild*. Somerville, Mass.: Boston Review; Cambridge, Mass.: MIT Press, 2006.

4 Corwin, Jeff. *100 Heartbeats: The Race to Save Earth's Most Endangered Species*. New York, New York: Rodale, 2009.

BIBLIOGRAPHY

Ayres, Ed. *God's Last Offer: Negotiating for a Sustainable Future.* New York, NY: Four Walls Eight Windows, 1999.

Balouet, Jean-Christophe, Eric Albert, and Joan Robb. *Extinct Species of the World.* New York: Barron's, 1990.

Barker, Rodney. *And the Waters Turned to Blood: The Ultimate Biological Threat.* New York, New York: Simon & Schuster, 1997.

Berry, Thomas. *The Great Work: Our Way into the Future.* New York: Bell Tower, 1999.

Brower, David and Steve Chapple. *Let the Mountains Talk, Let the Rivers Run: A Call to Those Who Would Save the Earth.* San Francisco, CA: HarperCollins West, 1995.

Buchmann, Stephen and Gary Nabhan. *The Forgotten Pollinators.* Washington, D.C.: Island Press, 1996.

Butler, Tom. *Wild Earth: Wild Ideas for a World Out Of Balance.* Minneapolis, MN: Milkweed Editions, 2002.

Caldicott, Helen. *Nuclear Power Is Not the Answer.* New York, NY: The New Press, 2006.

Campbell, Joseph and Bill Moyers. *The Power of Myth.* New York: Doubleday, 1988.

Carns, Ted. *Off On Our Own: Living Off-Grid in Comfortable Independence.* Pittsburgh: St. Lynn's Press, 2011.

Chivian, Eric and Aaron Bernstein. *Sustaining Life: How Human Health Depends On Biodiversity.* Oxford, New York: Oxford University Press, 2008.

Corwin, Jeff. *100 Heartbeats: The Race to Save Earth's Most Endangered Species.* New York, New York: Rodale, 2009.

D'Antonio, Michael. *Mortal Sins: Sex, Crime, and the Era of Catholic Scandal.* New York: Thorndike Press, a part of Gale, Cengage Learning, 2013.

De Becker, Gavin. *The Gift of Fear: Survival Signals that Protect Us from Violence.* New York, New York: Dell Pub., 1998.

Deloria Jr., Vine. *We Talk, You Listen: New Tribes, New Turf.* New York: MacMillan, 1970.

Devall, Bill. *Clearcut: The Tragedy of Industrial Forestry.* San Francisco: SierraClub Books/Earth Island Press, 1994.

Devorkin, David H., and Robert W. Smith. *The Hubble Cosmos.* Washington, D.C.: National Geographic Society and Smithsonian Institution, 2015.

Diamond, Jared. *The Third Chimpanzee: The Evolution and Future of the Human Animal.* New York: HarperPerennial, 1992.

Dowd, Michael. *Thank God for Evolution: How the Marriage of Science and Religion Will Transform Your Life and Our World.* Tulsa, OK: Council Oak Books, 2007.

Dowie, Mark. *Losing Ground: American Environmentalism at the Close of the Twentieth Century.* Cambridge, Massachusetts: The MIT Press, 1996.

Dudley, William (Editor). *Biodiversity*. San Diego, CA: Greenhouse Press, 2002.

Ehrenfeld, David. *The Arrogance of Humanism*. New York: Oxford University Press, 1978.

Encyclopaedia Britannica, 15th Edition. *The Inquisition*. Vol. 6: 328:2b. Chicago: Encyclopaedia Britannica, 2007.

Faux, Geoffrey. *The Global Class War: How America's Bipartisan Elite Lost Our Future – and What It Will Take to Win It Back*. Hoboken, N.J.: Wiley, 2006.

Flannery, Tim. *Now or Never: Why We Must Act Now to End Climate Change and Create a Sustainable Future*. New York: Atlantic Monthly Press; Distributed by Publisher's Group West, 2009.

Foreman, Dave. *Rewilding North America: A Vision for Conservation in the 21st Century*. Washington: Island Press, 2004.

Fraser, Caroline. *Rewilding the World: Dispatches from the Conservation Revolution*. New York: Metropolitan Books, 2009.

Gage, Matilda Joslyn. *Woman, Church and State*. Amherst, New York: Humanity Books, 2002. Originally published by C.H. Kerr: Chicago, Illinois, 1893.

Glavin, Terry. *The Sixth Extinction: Journeys among the Lost and Left Behind*. New York, New York; Thomas Dunne Books/St. Martin's Press, 2007.

Goodall, Jane and Marc Bekoff. *The Ten Trusts: What We Must Do to Care for the Animals We Love*. San Francisco: Harper San Francisco, 2002.

Greenberg, Paul. *Four Fish: The Future of the Last Wild Food*. Waterville, ME: Thorndike Press, 2010.

Hammer, Richard. *The Vatican Connection: The Astonishing Account of a Billion-Dollar Counterfeit Stock Deal between the Mafia and the Catholic Church*. New York: Holt, Rinehart, and Winston, 1982.

Hedges, Chris. *The World As It Is: Dispatches on the Myth of Human Progress*. New York: Nation Books, 2010.

Holden, Joseph. *Physical Geography: the Basics*. New York: Routledge, 2011.

Hughes, J. Donald. *American Indian Ecology*. El Paso: Texas Western Press, 1983.

Jacobs, Lynn. *Waste of the West: Public Lands Ranching*. Tucson, AZ: L. Jacobs, 1991.

Jensen, Derrick. *Endgame: The Problem of Civilization, Vol. 1*. New York: Seven Stories, 2006.

Josephy Jr., Alvin. *Patriot Chiefs: A Chronicle of American Indian Leadership*. New York: Viking Press, 1961.

Klein, Naomi. *This Changes Everything: Capitalism vs. the Climate*. New York: Simon & Schuster, 2014.

Kolbert, Elizabeth. *The Sixth Extinction: An Unnatural History*. New York: Henry Holt and Company, 2014.

Korten, David. *When Corporations Rule the World*. West Hanford, Conn.: Kumarian Press; San Francisco, CA: Berrett Koehler Publications, 1995.

Kump, Lee, James Kasting, and Robert Crane. *The Earth System*. Upper Saddle River, N.J.: Pearson Prentice Hall, 2004.

Kurzweil, Ray. *The Age of Spiritual Machines: When Computers Exceed Human Intelligence*. New York: Penguin, 2000.

Lambin, Eric. *The Middle Path: Avoiding Environmental Catastrophe*. Chicago University Press, 2007.

Laurance, William. *Stinging Trees and Wait-a-Whiles: Confessions of a Rainforest Biologist*. Chicago, IL: University of Chicago Press, 2000.

Leopold, Aldo. *A Sand County Almanac*. New York: Oxford University Press, 1949.

Liddick, Donald. *Crimes Against Nature: Illegal Industries and the Global Environment*. Santa Barbara, CA: Praeger, 2011.

London, Mark and Brian Kelly. *The Last Forest: The Amazon in the Age of Globalization*. New York: Random House, 2007.

Lovelock, James. *The Vanishing Face of Gaia: A Final Warning*. New York: Basic Books, 2009.

Lyman, Howard and Glen Merzer. *Mad Cowboy: Plain Truth from the Cattle Rancher Who Won't Eat Meat*. New York: Schribner, 1998.

MacDonald, Christine. *Green, Inc.: An Environmental Insider Reveals How a Good Cause Has Gone Bad*. Guilford, CT: Lyons Press, an imprint of The Globe Pequot Press, 2008.

Manes, Christopher. *Green Rage: Radical Environmentalism and the Unmaking of Civilization*. Boston: Little, Brown, 1990.

Margulis, Lynn and Karlene Schwartz. *Five Kingdoms: An Illustrated Guide to the Phyla of Life on Earth*. New York: W.H. Freeman, 1998.

Margulis, Lynn. *Symbiotic Planet: A New Look of Evolution*. New York: Basic Books, 1998.

Mason, Jim. *An Unnatural Order: Why We Are Destroying the Planet and Each Other*. New York: Continuum, 1997.

McDougall, Christopher. *Born To Run: A Hidden Tribe, Superathletes, and the Greatest Race the World Has Never Seen.* New York: Alfred A. Knopf, 2009.

McGavin, George. *Endangered: Wildlife on the Brink of Extinction.* Buffalo, N.Y.; Richmond Hill, Ont.: Firefly Books, 2006.

McKibben, Bill. *The End of Nature.* New York: Random House, 1989.

Meyer, Stephen. *The End of the Wild.* Somerville, Mass.: Boston Review; Cambridge, Mass.: MIT Press, 2006.

Mittenmeier, Russell. *Hotspots Revisited.* Mexico City, Mexico: Cemex, 2004.

Parenti, Christian. *Tropic of Chaos: Climate Change and the New Geography of Violence.* New York: Nation Books, 2011.

Pearce, Fred. *When the Rivers Run Dry: Water, the Defining Crisis of the Twenty-first Century.* Boston: Beacon Press, 2006.

Pearson, Richard and American Museum of Natural History. *Driven To Extinction: The Impact of Climate Change on Biodiversity.* New York: Sterling, 2011.

Pert, Candace. *Molecules of Emotion: Why You Feel the Way You Feel.* New York, New York: Simon & Schuster, Inc., 1997.

Posner, Gerald. *God's Bankers: A History of Money and Power at the Vatican.* NY, NY: Simon & Schuster, 2015.

Powell, James Lawrence. *Dead Pool: Lake Powell, Global Warming, and the Future of Water in the West.* Berkeley: University of California Press, 2008.

Preston, Richard. *The Wild Trees: A Story of Passion and Daring.* New York: Random House, 2007.

Quinn, Daniel. *Ishmael*. New York: Bantam/Turner Books, 1992.

Quinn, Daniel. *My Ishmael*. New York: Bantam Books, 1997.

Sapp, Jan. *What Is Natural? Coral Reef Crisis*. New York: Oxford University Press, 1999.

Shepard, Paul. *Man in the Landscape: A Historic View of the Esthetics of Nature*. College Station: Texas A&M University Press, 1991.

Shuster, Joseph. *Beyond Fossil Fuels: the Roadmap to Energy Independence by 2040*. Beaver Pond Press, Edina, Minnesota. 2008.

Smith, Laurence. *The World in 2050: Four Forces Shaping Civilization's Northern Future*. New York, N.Y.: Plume, 2011.

Suarez, Daniel. *Daemon*. New York: Dutton, 2009.

Suarez, Daniel. *Freedom™*. New York: Dutton, 2010.

Sussman, Art. *Dr. Art's Guide to Planet Earth: For Earthlings Ages 12 to 120*. White River Junction, VT: Chelsea Green Pub. Co., 2000.

Synnott, Mark. *Sins of the Aral Sea*. National Geographic, Vol. 227, No. 6, p. 114-131, January, 2015.

Tait, Malcolm. *Going, Going, Gone? Animals and Plants On the Brink of Extinction and How You Can Help*. London, England: Think Publishing; New York: Sterling Publishing Co. Inc., 2006.

Terborgh, John. *Requiem for Nature*. Washington, D.C.: Island Press, 1999.

Vaillant, John. *The Tiger: A True Story of Vengeance and Survival*. New York: Alfred A. Knopf, 2010.

Weintraub, David. *How Old Is the Universe?* Princeton, NJ: Princeton University Press, 2011.

Williams, Paul L. *The Vatican Exposed: Money, Murder, and the Mafia.* Amherst, New York: Prometheus Books, 2003.

Williams, Terry Tempest. *The Open Space of Democracy.* Great Barrington, MA: The Orion Society, 2004.

Wilson, E.O. *The Creation: An Appeal to Save Life on Earth.* New York: Norton, 2006.

Wuerthner, George. *Thrillcraft: The Environmental Consequences of Motorized Recreation.* White River Junction, VT: Chelsea Green Pub. Co., 2007.

Wuerthner, George. *Welfare Ranching: The Subsidized Destruction of the American West.* Washington, D.C.: Island Press, 2002.

WEBSITES USED IN REFERENCE NOTES ONLY

Associated Press. *"Dallas club gets permit to import dead rhino."* Arizona Republic, azcentral.com, March 29, 2015.

Bhat, Kiran. "Misplaced Priorities: Ethanol Promotion and its Unintended Consequences." *Harvard International Review.* Vol. 30; pg. 30-4. Spring 2008.

biologicaldiversity.org/news/press_releases/2015/wildlife-services-04-13-2015.html

biology.duke.edu/bio265/jlp13/myco.php?t=nutrient

Chadwick, Douglas. *The Truth About Tongass.* National Geographic Magazine. July 2007. ngm.nationalgeographic.com/2007/07/tongass/chadwick-text/7

Christy, Bryan. *Ivory Worship.* National Geographic Magazine. October 2012. ngm.nationalgeographic.com/2012/10/ivory/christy-text.

en.wikipedia.org/wiki/Lists_of_IUCN_Red_List_Critically_Endangered_species. (Easier to identify.)

enenews.com/ap-biggest-question-fukushimas-melted-fuel-asahi-fuel-mainichi-one-fuel-experts-nuclear-cores-hit-groundwater-could-melted-ground-world-never-case-like-fukushima-fuel-melted-fell-video. June, 23, 2015.

evolutionarychristianity.org

Ferris, Timothy. *First Glimpse at the Hidden Cosmos.* National Geographic, Vol. 227, No. 1, p. 108-123, January, 2015. http://ngm.nationalgeographic.com/2015/01/hidden-cosmos/ferris-text.

footprintnetwork.org/en/index.php/GFN/living_planet_report2/

Franzen, Jonathan. *Last Song for Migrating Birds.* National Geographic, July, 2013. http://ngm.nationalgeographic.com/2013/07/songbird-migration/franzen-text

Gouveia, Luisa. "Neochloris Oleabundans UTEX #1185: A Suitable Renewable Lipid Source for Biofuel Production." *Journal of Industrial Microbiological Biotechnology.* Vol. 36; pg. 821-26. 18 April 2009.

grimwade.biochem.unimelb.edu.au/cone/main.html

iucnredlist.org

Johnson, Robbin, and Ford Runge. "Ethanol: Train Wreck Ahead?" *Issues in Science and Technology.* Vol. 24; pg. 25-30. Fall 2007.

lakepowell.water-data.com

Lane, Jim. "Jet Stream: Biofuels Digest Special Report on Aviation Biofuels." *biofuelsdigest*. Word Press. 3 March 2010.

Maisels, Fiona, S. Strindberg, S. Blake, G. Wittemyer, J. Hart, E.A. Williamson. *"Devastating Decline of Forest Elephants in Central Africa."* Plos ONE 8, no. 3 (March 2013): 1-13.

mead.uslakes.info/Level.asp

ncdc.noaa.gov

Oliveira, Ana Christina, and Luisa Gouveia. "Microalgae as a Raw Material for Biofuels Production." *Journal for Industrial Microbiological Biotechnology*. Vol. 36; Pg. 269-74. 4 November 2008.

Pimentel, David. "Corn Ethanol as Energy: The Case Against US Production." *Harvard International Review*. Vol. 31; pg 50-3. Summer 2009. Opposing Viewpoints Resource Center.

Potera, Carol. "Corn Ethanol Goal Revives Dead Zone Concerns." *Environmental Health Perspectives*. Vol. 116; pg. 242-3. June 2008.

teslaenergy.com

Tyner, Wallace. "The US Biofuels Boom: Its Origins, Current Status, and Future Prospects." *Bioscience*. Vol. 58, Iss. 7; pg. 646-74. July/August 2008.

Wolinsky, Howard. "The Economics of Biofuels." *European Molecular Biology Organization Reports*. Vol. 10. No. 6; pg. 551-53. 2009.

BIBLIOGRAPHY, VIDEOS

Bidder 70. Gage, Beth and George.Telluride, CO: Gage & Gage Productions; [S.I.]: Distributed by First Run Features, 2013. DVD.

Chasing Ice. Orlowski, Jeff. New York: Docurama, Cinedigm Entertainment, 2013. DVD.

Cosmos. Sagan, Carl, Ann Duryan, Steve Soter, and Adrian Malone. [Volume 2], [Volume 2]. Studio City, CA: Cosmos Studios, 2000. DVD.

Cosmos: A Spacetime Odyssey. Duryan, Ann, Seth McFarlane, Mitchell Cannold, Brannon Braga, Jason Clark, Neil deGrasse Tyson, and Alan Silvestri. Cosmos Studios, Inc.; Fuzzy Door Productions, 2013. Beverly Hills, CA: Twentieth Century Fox Home Entertainment, 2014. DVD.

The Cove. Psihoyos, Louie. Santa Monica, CA: Lion's Gate Entertainment, 2009. DVD.

Cowspiracy. Kip Anderson and Keegan Kuhn. A.U.M. Films & First Spark Media, 2014. DVD.

Deliver Us from Evil. Berg, Amy. Santa Monica, CA: Lion's Gate Entertainment, 2006. DVD.

Emptying the Skies. Kass, Douglas and Roger. Chicago: Music Box Films, 2013. DVD.

The End of Poverty? Portello, Beth, Phillippe Diaz, and Martin Sheen. [Canoga Park], CA: Cinema Libre Studio, 2010. DVD.

End of the Line. Rupert Murray and Ted Dansen, Claire Lewis, George Duffield, and Charles Clover. New York: Docurama, 2009. DVD.

Evolutionary Christianity. Michael Dowd. Evolutionarychristianity.org, 2011-2015. DVD.

Fuel. Dir. Josh Tickell. Open Pictures. Blue Water Entertainment. 2008. DVD.

Gas Hole. Wagener, Jeremy and Scott D. Roberts. Burbank, CA: Cinema Libre Studios, 2008. DVD.

Gasland Part II. Josh Fox. New York: Docurama, Cinedigm Entertainment, 2014. DVD.

The Internet's Own Boy: The Story of Aaron Swartz. Knappenberger, Brian. Los Angeles, CA: Luminant Media, 2014. DVD.

League of Denial; the NFL's Concussion Crisis. Kirk, Michael and Mike Wiser, Steve Fainaru, and Mark Fainaru-Wada. Kirk Doc. Group PBS, 2013. DVD.

Racing Extinction. Psihoyos, Louie. Seattle, WA: Vulcan Productions, 2015. DVD.

Thrive: What On Earth Will It Take? Soquel, CA: Clear Compas Media, 2011. DVD.

Who Bombed Judi Bari? Mary Thomson, Liz and Darryl Cheney. Hokey Pokey Productions, 2012. Audio Book or DVD.

Who Killed the Electric Car? Culver City, CA: Sony Pictures, 2006. DVD.

ACKNOWLEDGEMENTS

My decades-long dream unexpectedly came true after I was laid off from a career that was no longer in demand; as a result, I found a U.S. government program which allowed me to go to college for free, including living expenses. I also have considerable gratitude for all those who led to my receiving the Washington Award for Vocational Excellence (WAVE Award) from Governor Chris Gregiore, especially Karen Halpern, the life-changing college instructor who recognized my potential and encouraged me to apply, the winning of which allowed me to continue my dream and receive a Baccalaureate Degree in Writing. I thank Donald Rhodes of Heritage Bank in Olympia, WA, for contributing my first scholarship, which changed my perspective about my possibilities. My sincere appreciation to those unsung heroes, my college writing instructors: Stephanie Groeschell, Nick Owen, Alex Gouirand, Peter Bacho, librarian Sarah Kaip, and also college executive director, Cecelia Loveless. My deep gratitude to Soma Clifton for inviting me to take care of his future retirement home, giving me the time and space to write without financial pressure; Jessica Leigh for offering me to read *Eco Barons*, by Edward Humes, the timing of which ignited my fire; the many brave, expert, and pioneering authors cited in this work; and all the Native American friends, teachers, ceremonialists, authors, and many tribes over the years. My deep gratitude extends to all those who have come before, such as the likes of David Brower, who had the foresight, love, and drive to succeed in preserving what gorgeous wilderness areas we have, and to all those who are similarly working to preserve the wild species of our Earth from extinction. And, of course, I thank Spirit, the Great Mystery, for your guidance and alignment of affairs.

INDEX

concentrated solar thermal power
147
Coriolus effect 61, 62
corporate Darwinism 33
Council of Nicaea 220
Cretaceous extinction 50
cryosphere 175, 176, 177

D

Darius 97
Dark Ages 221, 222
Darwinism 33
DDT 7, 8, 254
debt 11, 117, 118, 130, 143, 146,
    158, 230, 255, 256, 281, 282,
    299
DeChristopher, Tim 280
demographic winter 46
depleted uranium 157, 158, 165
derivatives 118
Descartes, Rene 29
Devonian extinction 49
dominion 26, 30, 33, 222, 257,
    259, 266

E

Ebola 84, 169
economies of scale 114
Edict of Milan 221
electric cars 150
Emerson 89
Endangered Species Act 12, 90,
    102, 109, 132, 204, 213, 278,
    298
endorphins 31, 40
Eocene epoch 83
ethanol 115, 149, 150, 151, 152,
    153, 154, 156, 168, 286, 289
ETs 169, 290
eukaryotes 39, 251
exosphere 65
externalities 111, 115, 116

F

farm salmon 189
feller-bunchers 77, 114
fractional banking 98, 117, 212,
    238
free market 111, 115, 118, 213,
    248, 274
Fukushima 149, 160, 185, 190, 233
Fuller, Buckminster 24

G

galactic year (GY) 22
Genesis 28, 30, 31, 32, 74
Ghost Demand Syndrome 57, 74,
    190, 268, 295
gibbons 9, 54, 83, 84, 85, 256
Glass-Steagall Act 131
Global Amphibian Decline 3
glyphosate 8
GMO 11, 77, 85, 98, 107, 214,
    248, 289
Goddess of the Yangtze 51
Golden Age of Discovery 267
Gomorrah 239
Gondwana 62, 175
Green Cross 287
green revolution 155, 246

H

Hadean eon 35, 51
Hadley Cell 62, 180, 183
Hanford 149, 157
HEE 9, 10, 11, 46, 50, 97, 155, 295
Hitler 236
Holocene geological epoch 10
Homo habilus 23, 123, 124
honeybees 127, 252, 253
howler monkeys 81
Hubble 21, 24
Human Exponential Event (HEE) 9
humanism 32, 33

paper parks 268
paradise xi, 16, 34, 57, 223, 245, 257, 258, 260
pastor David Dowd 231
patriarchs 28, 201, 224, 228
penguins 9
peptides 40
Permian extinction 49
Pine Island Glacier 142, 176
Pleistocene epoch 50
plutocracy 51, 274
plutonium 51, 157, 158, 160
Pope 221, 226, 237, 239, 241, 293
precession of the equinox 176, 197
Primorye 76, 287
Prohibition 150
prokaryotes 19, 20, 23, 35, 36, 38, 53
public lands 91, 206, 207, 210, 211, 213
pyrolysis 156

R

radioactivity 133, 185, 188, 255
Reagan 72, 153, 246
redwoods 9, 72, 75, 76, 251
religious theories 198
Renaissance 33, 222
Rockefeller, John D. 148

S

Santa Clara County v. Southern Pacific Railroad 129
sex abuse 239
Silicon Valley 168
Singapore 108, 121
stratosphere 4, 39, 65, 74
sustainable development 47, 231
symbiogenesis 38

T

tar sands 155, 156
tectonic subduction 61, 64, 184

temperate rainforests 67, 76, 287
Tesla 147, 148, 149, 150, 153, 156, 160, 233, 285
thermocline 61
thermohaline 61, 63, 184. *See* also thermocline
thermosphere 65
Thoreau 89
Tongass 287, 293
Triassic extinction 50
Trojan War 138
trophic cascade 81, 192, 203, 217
troposphere 17, 65, 165

U

uranium 157, 159
USDA 26, 71, 91, 208, 213, 217

V

Vatican Bank 237, 238, 239
vegetable varieties 13
Venter, Craig 35
Vietnam 121, 193

W

West Africa 13, 112, 169, 267
West Antarctica 142
wetlands 11, 101, 133, 166, 195, 203, 209, 243, 251
wildlife corridors 92, 266, 275, 287, 290
Wilson cycle 61
Witch Prickers 225
World Bank 76, 117, 130, 131

Y

Yellowstone 70, 92, 181, 203, 206, 266
Young, Brigham 228

Z

zoos 163, 195, 255, 300

ABOUT THE AUTHOR

Paul Rauch, aka Sequoia, is the founder and president of The Foundation for the Survival of Species, a non-profit organization (survivalofspecies.org), which educates people about endangered species and the ongoing mass extinction. He is a life-long naturalist and an Outward Bound graduate, having spent over a decade of summers soloing along the trails of the virgin mountain wilderness areas of the Pacific Northwest's ancient temperate rainforests of his youth, scrambling and bivouacking off-trail as well. He later earned a BA in creative nonfiction writing from The Evergreen State College. Having focused on nontraditional healing and spirituality for decades, he currently works as a counselor in Northern Arizona, contributing to one of the top couple's healing retreats in the U.S., providing outdoor sessions in the Coconino National Forest for couples and individuals.